Smart Technologies in Healthcare

Smart Technologies in Healthcare

Editor

Bruno Bouchard

LIARA Laboratory

University of Quebec at Chicoutimi, Canada

CRC Press is an imprint of the
Taylor & Francis Group, an **informa** business
A SCIENCE PUBLISHERS BOOK

CRC Press
Taylor & Francis Group
6000 Broken Sound Parkway NW, Suite 300
Boca Raton, FL 33487-2742

Printed on acid-free paper
Version Date: 20170124

International Standard Book Number-13: 978-1-4987-2200-1 (Hardback)

Visit the Taylor & Francis Web site at
http://www.taylorandfrancis.com

and the CRC Press Web site at
http://www.crcpress.com

Preface

Assistive Technologies in Smart Environments can help, transform, and enhance the way people with disabilities manage their daily lives and activities that would otherwise be difficult or impossible for them to do. However, despite the developments of a lot of new assistive technologies, much need to be done, especially concerning the establishment of standards and guidelines for the field. Users' impairments and particularities are so diverse that implementing complex technological solutions—mandatory for user adaptation—represents a major challenge in terms of universal design. Also, despite the rapid evolution of the field in the last decade, it is still considered as an emerging one. In this context, the main objective of this book is to investigate the most recent solutions to problems occurring in various aspects of assistive technologies specifically applied to the domain of smart environments.

This volume contains chapters covering the main aspects in the field of assistive technologies. The introductory chapter aims to show the historical evolution of the notion of assistive technologies in smart environment and to present it in the current context. It also presents the new paradigm of *open innovation* constituting the key to the success of the most prolific research teams around the world. The following chapters aim to throw light on the state of the progress in fields, such as: human activity recognition, automatic learning of the user's profile, human perspective of the technologies, physiological and cognitive monitoring, context awareness, user interface, security aspects, etc.

Activity recognition and automatic learning of human activities represent a core issue for developing useful assistive technologies. To be able to help a user, the system must first need to know what he is doing and how, in order to provide an adapted assistance. That is why a complete chapter on how applying data mining in smart homes has been included in this book. This part of the book is the most technical aspect of the problem in regard to computer science. It is also one of the subjects that have been the researched and published.

The human perspective constitutes another key aspect of assistive technologies. However, it has largely been overlooked and only been treated superficially in most papers. Nevertheless, this is a fundamental element of assistive technology for the acceptance of the developed new devices by the users and the caregivers. Therefore, the technologies must be developed for the people, according to the human perspective. Technology must adapt itself to its user, not the other way around. For that reason, we decided to include a full chapter presenting the core elements of the human aspects, which are fundamental to the development of assistive technologies.

Context awareness refers to the ability of a system to capture, model and use specific information about the environment surrounding the system, such as location, time, user profile, etc. Using contextual information to adapt the assistive services to the user constitutes a formidable challenge. In this book, a comprehensive chapter on this issue has been included. This chapter also includes a real case study.

Interfaces are the link between the system and the user. Therefore, they must be easy to use, be intuitive and efficient. To address this aspect of the problem, a chapter has been added that presents a case study with senior people on the usage and design of user interfaces. The results of the study can be used as guidelines for creating good interfaces.

Physiological and cognitive health monitoring constitute a prolific and important field of assistive technologies. Achieving efficient and effective monitoring is, however, not so easy. Which device or sensor to use? How to use it? How to design the device to be accepted by the user? How to process and presents the data in order to be useful? To answer these questions, two chapters have been included in the book to cover aspects related to health monitoring.

Finally, the security perspective of the technology should not be neglected. This issue is generally overlooked, in more attention is paid to mostly neglected in published literature technical or human aspects of the technology. However, the security issues are essential if it is to be deployed. What about fault tolerance? What about reliability of the system? What about the decisions taken by the intelligent system that may put the user in danger? In this book, we have included a full chapter on a new approach to address this important issue based on the Discrete Controller Synthesis

(DCS). This new intuitive approach allows for formally ensuring the security aspect of an intelligent system.

This book will serve as a reference for researchers, practitioners and engineers that seek to have a complete portrait of the field of assistive technologies.

Contents

1

Ambient Smart Assistive Technologies

Challenges and Perspectives

Bruno Bouchard[1,2,*] and *Sébastien Gaboury*[2]

1. Introduction

Today, most Western countries face an unprecedented demographic crisis caused by accelerated ageing of its population (United Nations 2013). This is made worse by a lack of resources and shortage of qualified home-care workers. Senior citizens, many of whom suffer from the loss of autonomy caused by cognitive or physical disorders, or both, wish to remain at home as long as possible. Staying in the home is clearly desirable not only from an economic point of view (Oderandi et al. 2012), but also because it offers a better quality of life by allowing the deinstitutionalization that is consistent with societal values: people should live as normal a life as possible without segregation and enjoy a dignified existence with full access to autonomy. For many seniors with moderate to severe functional dependence, however, ageing at home entails coping with numerous risks and practical challenges. The home environment has to be adapted, if not technologically enhanced, using intelligent technologies and sensors to offset cognitive and physical

[1] Intelligent Technologies Consultant (ITC enr.). Email: Bruno.Bouchard.PhD@ieee.org
[2] Laboratoire LIARA, Université du Québec à Chicoutimi, Chicoutimi, Québec G7H 2B1, Canada. Email: Sebastien.Gaboury@uqac.ca
* Corresponding author

deficiencies, to provide assistance and guidance, to ensure safety and to support natural caregivers and professionals in their work. This vision of the future, which has now become a reality, originated in 1988 at the Xerox Palo Alto Research Center (PARC), resulting in the work entitled *'The Computer for the 21st Century'* by American scientist, Mark Weiser (Weiser 1991). From the early 1990s, a large community of scientists developed around this specific research niche (Blackman et al. 2015), actively seeking technological solutions for these very human problems by employing such concepts as ubiquitous sensors, ambient intelligence (AmI) and assistive technologies to keep people in their homes. The idea achieved maturity in the mid-2000s and we have since seen the fruits of this research develop into tangible innovations, technologies to promote autonomy and quite substantial economic spinoffs. For instance, in the United States alone, the market for assistive technologies based on ambient intelligence will reach approximately $60 billion (US) a year by 2018, with expected annual growth of nearly 6 per cent over the next decade (BCC Research 2013).

1.1 What is Ambient Assisted Technologies?

In general, Ambient Assisted Technologies (AAT) refer to the use of an array of electronic devices—sensors and actuators or effectors—incorporated into everyday objects, such as cabinet doors, stoves, lamps, screens and so on in a transparent way, meaning that they are not visible to the user, in order to monitor the user's status and provide assistance as needed, such as advice, feedback, guidance or warning—of a stove left on, for example, based on information collected and historical data (Ramos et al. 2008). In the scientific literature (Queirós et al. 2015), the interdisciplinary challenges related to this area are regarded as hugely complex and many key questions require investigation. For instance:

- What technologies should be developed and applied in order to meet the needs of the people in need, natural caregivers and practitioners?
- What kinds of sensors and actuators are better used?
- How should the necessary models of artificial intelligence be developed in order to implement these technologies?
- How are they to be adapted to user profiles?
- How are they to be deployed and maintained?
- Can technology improve the ability of the people with diminished autonomy to perform day-to-day activities?
- What new skills will practitioners need to acquire?
- What are the productivity gains for the healthcare system?

- How to develop marketing strategies and ensure the transfer of technology to enterprises?
- What is ethically acceptable and what is not?

The great difficulty in addressing the scientific issues in a field where the needs are so pressing is attributed to a number of factors—the almost unprecedented interdisciplinarity of the questions to be researched, the substantial infrastructure and equipment requirements for prototyping, the difficulties research teams face in establishing partnerships with the public and private sectors and with users in order to carry out experiments that demonstrate the effectiveness of the proposed technological solutions and so on. A few teams actually have access to this combination of key factors, which are the basis for efficiency in developing ambient assisted technologies.

1.2 Assistive Technologies **vs.** *Automation*

Often, the term 'assistive technology' is used to describe what is, in fact, an 'automated system'. For the sake of this book, we think it is important to clearly define both the concepts.

- **Automated System.** A system made of sensors and actuators which perform actions on behalf of the person. Most of the time, this kind of system is made for carrying out the task for a person in order to alleviate his workload. It is good in the context of a factory, for instance, where you want to execute a maximum of tasks without human interventions, in order to minimize the number of employees. In the context of assistance, this is actually bad because the system fails to increase the user's autonomy.
- **Assistive System.** An assistive system is also made of sensors and actuators, like the automated one. However, we can say that the assistive system is smarter than the automated one. In this sense, instead of simply performing the task for the user, it will try to provide real-time assistance for completing activities to increase the autonomy of the person. Of course, in case of immediate danger, the system will perform actions directly, but the main purpose of the system is to help the user to carry out his activities; not to perform them on his behalf. This provides physical and psychological benefits to the user.

The difference between both the concepts may seem subtle. In fact, an assistive system is a specific form, or a more evolved form of automated systems. To be sure to understand correctly, let's conclude this section with a small example. Let's say that Peter, a cognitively impaired user with head trauma, is cooking a chicken on the stove. After 40 minutes, Peter is distracted by the phone and after the call; he goes to his bedroom and forgets the chicken. After a certain delay, let's say 20 more minutes, the smart home system will take some action. The question is: Which action should the system make? In the case of an automated system, it will simply cut the power of the stove. This action is correct in order to preserve the safety of the person, but it does little to help the user in his rehabilitation process of restoring his cognitive abilities. On the other hand, an assistive system would begin by sending cues, hints and reminders to the user. For instance, it could begin by flashing lights in order to draw a path to the kitchen. Once the user is in the kitchen, it would send an audio message telling him that the chicken on the stove is ready. A screen can also show a video example of how to get the chicken out of the stove and how to turn off the power. Of course, if the user is unresponsive to a reasonable amount of prompting, or if there is an immediate danger, the system will cut off the power by itself.

2. Assistive Technologies: Open Innovation Model

Developing assistive technologies is not a simple task, especially in the academic field. As we described in Section 1, the challenge is complex because it is hugely interdisciplinary. Most of the universities tend to separate each field of research in different faculties, departments, etc. This approach by fields of research tends to complicate interdisciplinary projects. Moreover, the funding agencies also tend to finance projects by specific field of research. For instance, in Canada, we have a specific national funding agency which supports research projects in the field of engineering and natural sciences. We also have another agency supporting projects in the medical field. But what about us? We are developing assistive systems with sensors and actuators (engineering), with artificial intelligence (computer sciences), which are adapted and tested with cognitively impaired patients (medical field). Another issue is related to the very applicative finality of what we are developing. Assistive technologies are not just concepts; they need to be conceptualized, prototyped and tested with real patients. To be efficient, a lab working in the field needs to be able to cover all those aspects properly. In the last decade, we saw a lot of research done in the field, but only few real assistive products came out of that good research. Often, research teams will only concentrate their efforts on the concept, or on the prototype. Teams often have difficulty in accessing real data, or having access to real users/patients to test their technologies.

To solve these issues, we deployed in our lab an innovative integrated research program covering the entire vertical process of technological development, using the 'Open Innovation' model developed by the renowned Harvard Business School (Cherbourg 2006) and recently adapted for assistive technologies by the British Academy of Management (Oderandi et al. 2012). In that model (Fig. 1), every technology generated by research must go through seven key steps in synergy with all the partners: users, practitioners, researchers, public agencies and private enterprises:

1. Pre-technology (for example, identification of needs)
2. Theoretical bases related to technology (for instance, data mining approaches)
3. Systems development and prototyping
4. Controlled-environment testing (laboratory, ideally with real infrastructure)
5. *In vivo* experiments (real context, with users, on a long period)
6. Promotion, intellectual property and technological transfer
7. Marketing, commercialization and creation of spinoffs and start-up

Obviously, while this area of research is highly interdisciplinary, its gravitational centre and scientific driver is firmly planted in Ambient Intelligence (AmI) (Ramos et al. 2008). The technology—sensors and actuators—required to set up such assistive systems already exists (Bouchard et al. 2014). The scientific challenge is to exploit it successfully using innovative approaches in AmI in order to create intelligent technologies that provide the needed services to users.

3. Conclusion

In conclusion, we can say that the field of assistive technologies is currently one of the most important research and development areas of the future. This leading-edge sector target four main objectives: (1) to generate benefits for society and individuals by preserving autonomy and quality of life for people with cognitive or physical disorders and their natural caregivers through the creation of assistive technologies that can be efficiently integrated within the living environment (home or seniors' residence); (2) to achieve scientific advances in the form of formal models, artificial intelligence algorithms, rehabilitation practices based on conclusive data, changes in working methods and so on, through the rigorous *in vivo* testing and development of the aforesaid assistive technologies; (3) to build bridges between research, practice, users and industry by bringing

Phase 1 Pre-technology	Phase 2 Formal and conceptual foundation	Phase 3 Development and prototype	Phase 4 Test in controlled environment (lab)	Phase 5 Representative experiment *in vivo* (in real context)	Phase 6 Technology valorisation and transfer	Phase 7 Commercialisation, startup, marketing
Progression						
• Identification of patients' needs • Evaluation of cognitive and physical profiles of targeted users • Design constraints for technology • Cost assessment of technological • Market study • Review and comparative analysis of existing technologies and related works	• AI models • Algorithm of signal analysis • Learning, datamining • Activity recognition • Guidance algorithms • Etc.	• Implementation • Prototyping real-size experimental systems • Validation of formal models with prototype	• Deployment in controlled environment • Experiments, simulation with recorded scenarios • Test with a small number of targeted users • Confrontation with practitioners • Results analysis • Efficiency • Robustness	• Deployment in real context (*in vivo*) • Experiments with targeted users (long period, sufficient number of users) • Efficiency and potential demonstration • Assessment of the needed changes of practitioners work	• Economical evaluation • Market analysis • Formation of practitioners • Partnership with companies • Technological transfer techno. • R&D with companies • I.P. Protection • Patents	• Evaluation of commercial success • Market strategy • Integration of technology in everyday life • Spin-off and start-up creation with university students

Fig. 1. Open innovation model applied to the field of assistive technologies.

the necessary disciplines together in a single location in order to pursue interdisciplinary research based on a participatory approach whereby users and practitioners can be involved throughout the process; and (4) to support the industry and economical growth by conducting technology transfer activities through the establishment of industrial partnerships and an enterprise spinoff strategy. To be able to do that, a research team needs to meet several requirements. First, the team needs to be interdisciplinary. For instance, our team is composed of researchers in the fields of: computer sciences, mathematics, engineering, occupational therapy, physiotherapy, geriatric, neuropsychology, education and economy. Secondly, the team needs to cover the entire Open Innovation model, which means having lab infrastructure to prototype, having access to real *in vivo* testing area with targeted users (for instance, a retirement home), etc. Finally, the team will need to have a solid network of partners: companies in the fields of technologies, enterprises that will be technology buyers, companies that will support the technologies, organisms and foundations, users associations, medical center, etc. Most importantly, all these partners must be present during all the phases of the development process, from pre-technologies to final commercialization.

Keywords: Aging Population, Smart Assistive Technologies, Artificial Intelligence, Open Innovation Model

References

BCC Research. 2013. Disabled and Elderly Assistive Technologies, Market Report.

Blackman, S. et al. 2015. Ambient assisted living tech. for aging well: A review. IEEE Intelligent Systems.

Bouchard, B. et al. 2014. A Smart Range Helping Cognitively-Impaired Persons Cooking, Proceedings of the Twenty-Sixth Annual Conference on Innovative Applications of Artificial Intelligence (IAAI-14), Association for the Advancement of Artificial Intelligence (AAAI) publisher, July 27–31, Québec City, Quebec, Canada. pp. 1–6.

Chesbrough, H. 2006. Open Innovation: New Imperative for Creating and Profiting from Technology, Harvard Press.

Oderanti, F.O. et al. 2012. Commercialisation of Assisted Living Technologies and Services: A Framework for Sustainable Business Models, The British Academy of Management (BAM), Cardiff, UK.

Queirós, A. et al. 2015. Usability, accessibility and ambient-assisted living: A literature review. Journal Universal Access in the Information Society, Springer. 14(1): 57–66.

Ramos, C. et al. 2008. Ambient intelligence—the next step for AI. IEEE Intelligent Systems. 23(2): 15–18.

United Nations. 2013. World Population Ageing. Dep. of Economic and Social Affairs: Population Division. p. 114.

Weiser, M. 1991. The Computer for the 21st Century, Scientific American. 265(3): 94–104.

2

Challenges in Developing Smart Homes

Human Perspective

Julie Bouchard,[1,*] *Miryam Lépine El Maaroufi*[2] and *Marie-Pier Dufour*[2]

1. Introduction

Smart homes (SH) using artificial intelligence (IA) are a new housing model that allows cognitively impaired individuals to lead independent lives while providing temporary relief to their families and caregivers from their obligations. Even in everyone's best interest, many aspects—prompting adaptability, people with cognitive deficit capacity to use them and acceptability—are to be taken into consideration in the development and implementation of new technologies in a patient's familiar environment and, more particularly, when one is cognitively impaired. Results from a study conducted on these requirements are presented in this paper. The following chapter will be devoted to two of these man's perspectives aspects—adaptability and acceptability.

[1] DSS, UQAC, 555 blvd. Université, Chicoutimi, Québec, G7H 2B1, Canada. Tel: 418-545-5011 #5667. Email: julie1_bouchard@uqac.ca

[2] DSS, UQAC, 555 blvd. Université, Chicoutimi, Québec, G7H 2B1, Canada Tel: 418 545-5011 #2355.

* Corresponding author

2. Adaptability

2.1 Daily-living Activities

The new technologies for smart homes usually provide independence to the cognitive-affected person by assisting him in the performance of his everyday tasks. These tasks, called activities of daily living (ADL), can be divided in two main categories—personal (or basic) ADL (BADL), or domestic and community ADL (or instrumental) (IADL). It has been suggested that BADL "is typically restricted to activities involving functional mobility (ambulation, wheelchair mobility, bed mobility and transfers) and personal care (feeding, hygiene, toileting, bathing and dressing). As for IADL, it is often referred to as a series of life functions necessary for maintaining a person in his or her immediate environment, namely; money management, shopping, using the telephone, travelling within community, housekeeping, preparing meals and taking medications as prescribed (James 2008). These systems are usually developed to assist cognitively-impaired people in the performance of one of the BADL or IADL, but they can rarely assist them in several tasks simultaneously.

2.2 Prompting

Smart-home systems use various types of technology—some can detect the resident's error and provide guidance in these moments exclusively (either by video, step-of-task detection, or movement detection), some respond to the resident's request (voice request, touch of a button) and others provide help automatically (Boger and Mihalidis 2011). Some technologies take action for the resident; others only give hints so that the person can perform the task on his or her own (stimulating the person to maintain certain independence). The latter uses prompts in order to assist the person in the performance of tasks of everyday life (IADL or BADL) but not all prompts can be equally advantageous. When demonstrating the general efficacy of a prompting device, the current way is to compare it with that of human assistance (human versus technological efficacy) (Mihalidis et al. 2008) but, efficacy is not demonstrated for all devices.

Even if most smart homes use voice prompting to support the user, this type of prompting is not fully adapted for all kinds of cognitive disorders (Van Tassel et al. 2011). For example, verbal prompts might be more difficult to understand for people with language deficits since the perception and understanding of such prompts is compromised by their cognitive condition, even though they are overused in smart homes. Several conditions, for example, certain types of dementia, affect communication skills, making it difficult for users who can hardly express their needs to developers and caregivers to make assistance optimal and efficient and thus, enhance psychological well-being.

Several types of prompts can be used—some have already been tested for effectiveness although they do not correspond to the capacities and deficits of those they are supposed to provide assistance for. Many devices or systems have also been developed for different cognitive problems presented by patients (Boger and Mihalidis 2011). The need to focus on finding interesting prompting alternatives for smart homes is a challenge in this field.

Consequently, several impairments caused by ageing or disease, such as visual and auditory decline, loss of tactile sensation, motor disabilities, equilibrium disorders and cognitive impairment (Hoey et al. 2010) might have a negative impact on their abilities to use technological devices. For example, text prompts, which require that the user still has intact reading and direction of attention, would not be effective for people with important visual or severe cognitive loss because they are unable to read the prompts (Evans 2003; Gitlin et al. 2002).

2.3 Higher Cognitive Demand

Despite the fact that automatic assistance systems are designed to reduce cognitive burdens, they can often be cognitively demanding, especially for a patient who is already facing decreasing cognitive skills. This emphasizes the need for systems' obedience for residents experiencing severe cognitive disability (LoPresti et al. 2004). For this reason, choosing the right type of prompt is relevant to solve these issues. Presenting a prompt using the same sensory modality as the task requested might generate an interruption of its progress. For example, a video or photo prompt could be a distraction when the visual modality is needed to accomplish the task (Wherton and Monk 2008).

While verbal prompts mostly used in smart homes (Van Tassel et al. 2011) involve language comprehension, they may pose an additional challenge that will intensify cognitive burden (Wherton and Monk 2008). Considering the fact that a few studies discuss the effectiveness of the different kinds of prompts, it has been observed that prompts produce best results if they are tailored to individual preferences (Boger and Mihalidis 2011). Some prompts may be very helpful for a few people while proving irritating or confusing for others (Carrillo et al. 2009). They have to indemnify the users' insufficiencies by taking advantage of the person's remaining capacities in order to be effective (Van Tassel et al. 2011).

It is also essential to consider individual characteristics and adapt assistance technologies according to the patient's functional level (Aanesen et al. 2011). The biggest challenge to technical aspects of automatic assistance strategies is to discover the right combination of flexibility and simplicity (Carrillo et al. 2009).

2.4 Emotional Reaction to Prompt

Moreover, patients' emotional and behavioral reactions to certain types of prompts must be taken into account. In fact, since they are aware of their decreased cognitive functions, patients become more anxious. For instance, some think they are slowly losing their minds and sending a vocal prompt may lead them to think they are 'hearing voices', which may increase their anxiety and disorientation (Van Tassel et al. 2011).

Devices and prompts are thus developed to respond to either cognitive deficits encountered (memory loss, orientation, language deficits, etc.) or to compensate performance of some aspects of everyday life (mobility impairments, feeding and hearing loss, etc.). Many of them are already in use or tested for research purposes.

2.5 Examples of Devices Developed for Cognitive or Activity Purposes

There are various kinds of memory deficits. Among these, short-term memory loss not only has deleterious effects on daily-life autonomy but also affects IADL and BADL and the capacity to learn to use the technology. This symptom is encountered in many cognitive diseases; it causes learning difficulties even for the smallest needs, such as how to use a new gadget or think of taking the medication (Côté and Hottin 2012). The Independent LifeStyle Assistant™ (I.L.S.A.) was developed for patients with this kind of memory problem (Haigh et al. 2004) and may be useful for the cognitively-affected person and his caregivers. It has interesting features, such as monitoring, date reminders, alerts and notifications, summary report of the patient's behavior and much more (Haigh et al. 2004). With equipment, such as iPads and Android tablets, patients can see this information at all times and follow their schedules (Haigh et al. 2004). Caregivers can access this system through a wireless access connection and follow in real time the data collected by the system (Haigh et al. 2004). They can view and edit the prescriptions and alerts/reminders, and receive on their mobile phone the patient's abbreviated status reports, records and schedule reminders. In short, they can keep in touch anytime from anywhere (Haigh et al. 2004). For patients with cognitive problems (mostly with memory deficit), managing time and adjusting to the requirements of their environment on their own is a complex task. Because of the learning difficulties they experience, the developed interface of the devices must be user-friendly and enable patients to use it on their own (Imbeault et al. 2011). Other devices, such as connected homes, also allow caregivers to pay bills and do grocery-shopping online remotely (Imbeault et al. 2011). These patients can also benefit from telehealth—a system linking patients to doctors or any other health professional, allowing the doctors to follow patients' collected data in real time, or even videoconference with them, if needed (Gentry 2009).

In support of prospective memory difficulties (the inability to remember performance of future tasks), memory glasses have been developed to record events of the day (DeVaul et al. 2003). Through artificial intelligence, it takes note of the context and captures images to provide appropriate reminders, or procedural guidance in task execution, with a customizable multimodal display and prompts using text, audio or visual graphics adapted to the patient's remaining cognitive capacities (Bharucha et al. 2009). It can also help in mild cognitive impairment by providing names of people they encounter (anomia) and facial recognition (prosopagnosia) (Bharucha et al. 2009).

For retrospective memory impairments (when the content that is difficult to remember is in the past), Microsoft's SenseCam has developed a wearable digital camera that takes photographs while worn (Doherty et al. 2013). After recording the events of the day, the user can review these images to enhance consolidation of autobiographical memory (memory of episodes recollected from an individual's life and also of the knowledge of this person) (Bharucha et al. 2009).

Orientation difficulties can cause the person to be less familiar with his environment (Côté and Hottin 2012). At times, individuals may be so confused that they can eat breakfast in the middle of the night, search for their own bathrooms or even fail to recognize their own home. This is why several models of abnormal behavior patterns have been developed using wireless sensors that detect pressure and open doors automatically (Navarette et al. 2012). They detect if the person is lying on the ground, wandering too much (anxious crisis) or remaining in the same room for an excessive amount of time so that the system can contact caregivers for emergency (Navarette et al. 2012). These researchers decided to work with sensors since it was important for them to avoid cameras and microphones devices, which are unethical and invasive for the patient's privacy (Navarette et al. 2012). Furthermore, digital screens within the home are used to provide the Alzheimer patient with useful, easy-to-understand short messages or pictures. During the night, short messages, like 'it is night time, go to bed', can be seen on a digital clock (Donnelly et al. 2011). This system can command patients to get out of the bed at any time in the night. Digital screens can be installed in every room with a predictive behavior system that shows (with light) the way to the kitchen when it is lunchtime (Donnelly et al. 2011). There is also the ENABLE project which provides calendars within the home to show the patient the exact moment of the day (date, time, morning, evening, afternoon and night) (Holthe and Engedal 2004).

People with severe cognitive disorders are sometimes found wandering and get lost in their surroundings, even in the middle of the night if not under surveillance (Gentry 2009). A wearable device, such as an electronic watch with an integrated GPS, can direct patients if they leave the house (Wherton and Monk 2008); an arrow pointing to the right direction would be displayed on the screen of the watch for support (Wherton and Monk 2008). Again, bracelets or pendants could send distress calls only on the push of a button, which would alert the caregivers to react appropriately (Gentry 2009).

2.6 Recognition and Prediction Patterns

For every possible solution enumerated, it is important to be able to predict the behavior of these patients. This is why Donnelly et al. (2011) developed the Dante system—a system using objects tagged with a marker to interpret and monitor an occupant's behavior in order to predict it (Donnelly et al. 2011). The Georgia Institute of Technology is also developing the Aware Home—an experimental intelligent environment informed on itself as it is on its inhabitants (Kidd et al. 1999), which will be useful to recognize the resident's potential crises and behavioral trends and in preventing disorientation and crisis (Frisardi and Imbimbo 2011).

2.7 Mobility Impairments

Even with cognitive deficits, a patient can suffer from mobility impairments, making it difficult for him to adjust light intensity or reach unattainable switches in the house. This is why remote controls using frequency signals (such as Z-Wave or ZigBee), voice control or even-eye gaze control can help a resident to adjust the different units within the home in the simplest way possible (light, volume, temperature, etc.) (Gentry 2009).

Falls are also a major risk that must be accounted for when elderly people with neurological impairments are living alone at home. This is why, in recent years, detecting falls at home has become crucial. Rougier et al. (2011) developed their first project which makes people wear sensors that detect falls; many found these kinds of measures too intrusive. For this reason, the Rougier team developed a new project using external sensors, like floor vibration detectors and camera systems, to detect falls in accident and warn family caregivers (Rougier et al. 2011). This solution respects the elderly's privacy and is available at any time of the day.

2.8 Meals

Without the supervision of caregivers, many patients with cognitive deficits would not be able to eat properly and follow a healthy diet (Huang et al. 2010). They would forget to have their meals or what they had for dinner the previous night. Some researchers are actually working on a personal diet-suggestion system that registers human-related factors and provides information on what the patient needs to eat, depending on different contexts (for example, asking the resident whether it is breakfast or dinner, if he suffers from diabetes, on the iron levels in his blood, or his food preferences, etc.) (Huang et al. 2010). It would even take account of the specialist's recommendations (Huang et al. 2010). Moreover, patients suffering from diabetes could benefit from a personal diabetes-monitoring system (Zhou et al. 2010). For this system, wearable sensors are required to collect physical signs, like blood glucose level or blood pressure records that are then sent to a cell phone (Zhou et al. 2010). It records daily test results to provide follow-ups to family caregivers and health professionals at any time.

2.9 Language Impairment

For language-impaired patients, the VERA (The Visually Enhanced Recipe Application) provides visual cooking instructions along with full texts and sounds; they are customizable to accommodate to people individual's strength (Tee et al. 2005). Along the same lines, the Cook's Collage displays the previous steps of the cooking task with a captioned video whenever the user is lost in the different steps (Tran et al. 2007).

2.10 Hearing Loss

Some patients may suffer from hearing impairments as well. For this reason, it is imperative that the different warnings from these smart-home units are delivered to patients simultaneously on devices such as mobile/phone/tablet/wristwatch (Lozano et al. 2012) or in any other form, such as visual prompting (e.g. light flashing). They need to receive a message or a picture alerting them of any disruption within the environment (the doorbell ringing, someone calling, the oven still operational) (Lozano et al. 2012).

2.11 Healthcare

Video visits are a fast way for nurses and patients to communicate (Aanesen et al. 2011). A two-way video camera, where the patient and the nurse see each other in real time, enables the nurse to give instructions or simple tests, such as measuring sugar level in the blood and reporting back the results instantly (Aanesen et al. 2011). Patients with memory loss and chronic diseases need to be under constant surveillance; therefore, this device provides fast instructions to the patient and fast feedback to the healthcare clinician, thus reducing the costs of physical visits (Aanesen et al. 2011). Studies show that elderly people are satisfied with quick and simple information received through video visits (Aanesen et al. 2011).

2.12 Social Interactions

The CIRCA multimedia platform (Computer Interactive Reminiscence and Conversation Aid) was developed to ease communication between caregivers, family and patients with cognitive impairments, such as memory loss (Alm et al. 2004). People with dementia use the touch screen and select a reminiscence theme available in three media types (photographs, video or music), which allows them to choose the topic of the conversation (Alm et al. 2004). The data base and tracks available can be edited any time by the caregiver, who also has access to it (Topo et al. 2004). The Social Memory Aid is designed to improve the user's memory (Joseph and Monk 2008). It is a data base that contains different photos of people and provides information about the selected picture (Morris et al. 2004). Questions and clues are projected on the TV screen, making it possible to practice remembering details about friends and family (Joseph and Monk 2008).

2.13 Leisure

The Picture Gramophone (Topo et al. 2004) is another platform which offers music on a touch screen with lyrics displayed in a large text while music is being played (Joseph and Monk 2008).

3. Familiarity, Learning, Cognitive Deficits and Generalization

People with dementia tend to be more proficient when they are in familiar settings, but are more prone to making errors when moved outside of their usual surroundings (Wherton and Monk 2008). Although familiarity of the environment can only be achieved partially with the use of technology and because of the changes that technological interventions bring in daily activities, it is essential for designers to create prototypes as natural and unobtrusive as possible to ensure the effectiveness of the systems and the adaptation of users to them (Wherton and Monk 2008).

In addition, many cognitive deficits hinder learning of new skills that are needed to use the technology (e.g. anterograde memory). Learning processes can be affected by various cognitive troubles, affecting working and episodic memories before the activity becomes automated (procedural memory) (Beaunieux et al. 2006). However, long-time learned processes are more likely to resist (Beatty et al. 1994; Beatty et al. 1997). Inconveniently, for the most part of potential users, namely elderly people, digital technologies are not usual and can be quite intimidating because of their complexity. Since retaining new information is challenging for people with memory deficits (Cook and Das 2007), the time and efforts required to learn the use of technologies may be an important barrier for them and not everyone will be capable or ready to use them (Chan et al. 2009; Aanesen et al. 2011). This characteristic must be accounted for when developing technology for these people. Technology must thus be user friendly and require little learning. Certain pathologies will cause the cognitive deficits to increase with time (e.g. Alzheimer), thus necessitating systems to detect these changes and adapt to them. A third characteristic that is not thoroughly examined when developing technology for cognitive-impaired people is generalizability. When a technological device only assists in the performance of one task and many different devices are needed within the home to make it work, it can be very difficult for patients to adapt to these various devices. If such a device uses only one way of interacting and for different tasks, it may then be more practical for patients with cognitive deficits.

4. User's and Caregiver's Needs and Attitudes

One of the most challenging aspects to handle automatic assistance is the extremely large variability of target users' individual characteristics. Thus, variables, such as mental abilities, experience and backgrounds are unpredictable aspects that impede on operational modes and arrangements of smart-home equipment. Smart homes are meant to sustain older adults, people with chronic illnesses and physically or cognitively handicapped people living on their own at home (Chan et al. 2012). At first, assistive technologies were predominantly designed for younger people with non-progressive brain injuries; later, researchers attempted to generalize the prototypes to people suffering from progressive deficits caused by neurodegenerative dementia (Bharucha et al. 2009). These deficits are progressive in the sense that needs and capabilities of patients change over time; technologies need to be developed and adapted to such progression in order to constantly support the user through the evolution of his disease or condition (Hoey et al. 2010). Patients suffering from chronic illnesses and disabilities may also face unpredictable and temporary changes in their condition (Demiris et al. 2010), and may, as well, need the same kind of assistance from short-term to long-term. The systems implemented need to be constantly adjusted to the changing conditions and needs of the patients in order to perform well; they cannot be static and homogeneous for all patients, considering the important heterogeneity of the individual's profile. To summarize, it is necessary to adjust the technology to the patient's level of adaptation, his remaining strengths and particularities for these devices to be efficient and useful in home care. The industry has to take notice of the problems the patients are facing individually, consider their functional ability and the most appropriate assistance solution to choose before proposing impersonal and ineffective devices (Aanesen et al. 2011).

5. Individual Differences

Not only that intergroup features are very irregular, but they also are quite unsteady within groups of individuals with the same condition. For instance, dementias represent an important diversity in symptomatology; the most common manifestation is memory loss, especially for recent events and decrease of other cognitive functions, such as perception and comprehension of language. People with sustained acquired brain injury (ABI) are also a group of people that can benefit from smart homes and who present diversity in deficits (attentional executive, language, etc.). Two other groups targeted by the technology are older adults and people who have intellectual disability (ID). All of these groups present different profiles of capacities and deficits, and even within a group, they differ individually.

Many cognitive deficits, including attention-control processes, executive function or memory deficit may unsettle the ability to adopt unfamiliar measures (Wherton and Monk 2008), such as monitoring systems or electronic devices. In addition to the fact that elderly people are frequently uncomfortable with digital technologies because it is new or they are unacquainted, cognitive impairment emphasizes the difficulty to adapt to new procedures and environments (Wherton and Monk 2008). Even if people are able to communicate, cognitive deficits can impede on the daily use of computers, tablets and cell phones. This is why, to increase the usefulness of their products, some smart-home developers use add-ons that can read text a loud and display easy-to-understand pictures or messages on screens (Gentry 2009). They also use voice recognition and other simplified software to improve their capacities for using technologies (Gentry 2009).

Regardless of the variability in general personal characteristics, the needs are also different from one person to another. Therefore, it remains essential that healthcare technologies should not be set only by technological progress, but adjusted to individuals' motives and needs (Chan et al. 2009; Frisardi and Imbimbo 2011). As Frisardi and Imbimbo (2011) pointed out, technological development of cognitive aids must consider the needs and user's acceptance of the technology in order to be really effective. Moreover, they indicate that research aimed at demonstrating the possibility of delaying placement and long-term effect on independence when using IA are cruelly lacking.

In the same way, only a few experimentations, involving people with real impairments and disabilities, have been done, which makes it difficult for the industry to identify and understand what the customers really want, limiting their actual application to this population (Bharucha et al. 2009).

New ways are developing to test the technology with, for example, actors substituting real patient and simulating cognitive deficits in order to test the device.

5.1 Quality of Life

The success of assistance technologies should not be measured only by functional progression in a particular department; the considerable benefits on the user's quality of life should also be taken into account (Bharucha et al. 2009; Pilotto et al. 2011). This subjective parameter is difficult to evaluate because people with cognitive impairment also lose the capacity to make judgments and insights. Psychological well-being is one of the major components of the quality of life measurements which rely on different factors depending on the issues, whether it is dementia or cognitive deficit (Ready et al. 2003). Moreover, when physical functioning problems add to cognitive deficits, risks are increased with the use of certain types of technologies, which become dangerous and even unusable for people with physical deficits.

Again, disturbance of mood, changes in personality and behaviour may be a symptomatic aspect. They can complicate interventions and transformations of the familiar environment and make it challenging; patients can become upset because of technology installations. Drastic changes must be avoided since these may be the cause of disorientation in patients and bring additional challenges; for instance, they may lose the ability to recognize familiar objects or what they are meant for, or experience disorientation in time and space. Patients may get frustrated simply by having to use the new technology or due to the fact that it does not satisfy their needs (Walker et al. 1998; Van Tassel et al. 2011; Cavallo et al. 2016).

6. Acceptability

It is important for researchers to take into account the attitude and perception of the family and the caregivers in order to gain better acceptance of the technology. If the caregivers and the family feel safer and are relieved from the workload that taking care of the diseased person requires, it means that the technologies are useful and serve as expected (Rialle et al. 2008). For example, a study (Rialle et al. 2008) shows that the most appreciated of technologies is the tracking device, which increases the patient safety while decreasing the caregiver's fear of wandering or accidents (Rialle et al. 2008). It is followed by videoconferencing, allowing individuals to remain connected to their loved ones at all times (Rialle et al. 2008). Thus, these technologies must be adapted to the patients as they should be to their caregivers in order to be fully accepted.

7. Patient and Caregiver's Emotional Aspects Towards Technology

Poor self-esteem, low quality of life, social isolation and anxiety are related to dementia and to other cognitive deficits. They affect both people with the deficit and their caregivers (Tinetti et al. 1990; Burns and Rabins 2000). Although automatic-assistance technology is quite promising since it aims at alleviating these problems by improving autonomy and personal well-being, it doesn't seem to generate enthusiasm from the general population (Rialle 2007). In fact many people, especially older ones, do not think that it is actually possible to enhance well-being and quality of life by using health-related technology (Archer et al. 2014). Oddly enough, the ones who are receptive to assistive technologies in fictional scenarios tend to deny personal needs when automatic assistance is offered to them. Indeed, they don't perceive themselves as potential users in need of such help (Beach et al. 2009; Coughlin et al. 2007). Often, even after people accept to try it, maintaining long-term acceptability is challenging; the abandonment rate of any assistive technology is significantly high and there are a number of reasons that can explain this. The most important reasons given are that the device does not serve the specific need; it is too difficult to use; or it cannot be customized (Tsui and Yanco 2010). Due to the vulnerability of the population with cognitive impairment, prototypes are rarely tested on people with real needs, the potential users. It is very challenging for designers to truly direct their concept according to individual needs and wishes (Boger and Mihailidis 2011). The frustration generated from unsatisfied expectations can therefore, interrupt the technology development process. The patients are dependent and must have a close relationship with those who take care of them and of their environment to continue living at home and avoid placement in a care center. In fact, the user's apprehension from the impression sent to others upon his adoption of this technology is most likely to be the most decisive factor influencing the success or failure of it (Coughlin et al. 2007). Expressing the desire for such assistance might result in a feeling of weakness and infantilization for some, which prompts them to deny their need of it. The inherent stigmatization associated with their use and the fear that electronic components could substitute the soothing company of caregivers may lead to mistrust or even complete technophobia-inducing adoption issues (Carillo et al. 2009). If people think that automatic assistance shortens or excludes the contact with meaningful people, they will tend to reject it. Because smart homes hold the promise of care-burden, it is important to consider the emotional benefits of daily human contact and friendship provided by human caregivers (Gentry 2009). Important sensibility to anxiety and stress, which is one of the most frequent symptoms in dementia, may enhance the difficulties encountered with modification in user's lifestyle and daily activities (Selmès and Derouesné 2004).

8. Caregiver's Reaction

Presently, automatic assistance aims at supporting caregivers while decreasing their care burdens without wiping them out completely. Some might be feeling guilty about relying on such kind of help because they think it means that they are capitulating on their caregiver's role. On the contrary, some might be afraid of the complexity and time needed to implement technological devices and so be reluctant about it. Most of the devices proposed by the industry need the intervention of a caregiver to perform optimally, so caregiver's involvement is essential to make it beneficial and effective. Consequently, the caregiver's attitude is fundamental to ensure the successful adoption of assistive technology. This is why designers must involve them in the development and implementation of in-home devices, to increase adoption possibilities (Bharucha et al. 2009). Since perception of an assistive technology has a major influence on whether or not people will use it, acceptability should be considered carefully while evaluating their effectiveness (Scherer 2005; O'Neill and Gillespie 2008).

9. Interconnections Between Adaptability and Acceptability

It is essential to recognize that the two major aspects in the development of this technology are interconnected. In fact, if researchers take time to examine the real needs and the abilities of the people they are targeting as users, it may increase the benefits of developing the devices. Often, devices are developed without consulting the population on their needs, their fears about the technologies, their intended impact on their life and on enhancement of their quality of life. Taking time for these approaches can prove rewarding—taking account of the needs and fear of the users (increasing their acceptability) and helping in the development of more adapted and thus, more usable technology (adaptability). Lastly, one must remember that the more the technologies are used, the less they will cost, and the more accessible they will become to all.

10. Conclusion

Many elements need to be contemplated when developing new technology in order to be accepted and used by the targeted clientele.

First is the adaptability to the user's cognitive deficits, skills, limitations, learning capacities and needs. If technologies are not adapted specifically to their users, there is great risk that they may not get adopted, that they are used only for a short period of time or they are simply set aside. The use of one single system to answer multiple needs of one single person in a same pattern is an interesting path to ease the help process. Developing means that are not needed is also to proscribe. It might be a good idea to spend time with the person to establish his real needs before implementing an array of devices.

The second aspect, acceptability, is overlooked too often. In fact, using technologies can have a significant impact: for the patient, the perceived stigma experienced when using certain technology tools, the fear and resistance that come with its use, the denial of need for it (which implies the recognition of technology deterioration), the fear of losing the loved one who takes care of him or her during this difficult time. If all these emotional aspects are not addressed with people, there may be a form of sabotage of the technology effectiveness. Similarly, caregivers can react, even when they are already voicing loud their fatigue and need for help.

The other elements not addressed in this paper, such as cost of implementation, ease of use, changes according to the situation's evolution and reliability are crucial points to appraise in the development of such technology.

Keywords: Attitudes, cognitive deficits, familiarity, acceptability, adaptability, learning, needs, smart homes

References

Aanesen, M., A.T. Lotherington and F. Olsen. 2011. Smarter elder care? A cost-effectiveness analysis of implementing technology in elder care. Health Informatics Journal. 17(3): 161–172.

Alm, N., A. Astell, M. Ellis, R. Dye, G. Gowans and J. Campbell. 2004. A cognitive prosthesis and communication support for people with dementia. Neuropsychological Rehabilitation. 14: 117–134.

Alper, S. and S. Raharinirina. 2006. Assistive technology for individuals with disabilities: A review and synthesis of the literature. Journal of Special Education Technology. 21(2): 47.

Archer, N., K. Keshavjee, C. Demers and R. Lee. 2014. Online self-management interventions for chronically ill patients: cognitive impairment and technology issues. Int. J. Med. Inform. 4: 264–72.

Beach, S., R. Schulz, J. Downs, J. Matthews, B. Barron and K. Seelman. 2009. Disability, age, and informational privacy attitudes in quality of life technology applications: Results from a national web survey. ACM Transactions on Accessible Computing 2, Article 5. 5: 1-5: 21.

Beatty, W.W., P. Winn, R.L. Adams, E.W. Allen, D.A. Wilson, J.R. Prince, K.A. Olson, K. Dean and D. Littleford. 1994. Preserved cognitive skills in dementia of the Alzheimer type. Arch. Neuro. 21.51(10): 1040–1046.

Beatty, W.W., R.A. Brumback and J.P. Vonsattel. 1997. Autopsy-proven Alzheimer disease in a patient with dementia who retained musical skill in life. Arch. neurol. 54: 1448.

Beaunieux, H., V. Hubert, T. Witkowski, A.L. Pitel, S. Rossi, J.M. Danion and F. Eustache. 2006. Which processes are involved in cognitive procedural learning? Memory. 14(5): 521–539.

Bharucha, A.J., V. Anand, J. Forlizzi, M.A. Dew, C.F. Reynolds, S. Stevens and H. Wactlar. 2009. Intelligent assistive technology applications to dementia care: current capabilities, limitations, and future challenges. The American Journal of Geriatric Psychiatry: Official Journal of the American Association for Geriatric Psychiatry. 17(2): 88–104.

Boger, J. and A. Mihailidis. 2011. The future of intelligent assistive technologies for cognition: Devices under development to support independent living and aging-with-choice. NeuroRehabilitation. 28(3): 271–80.

Burns, A. and P. Rabins. 2000. Carer burden in dementia. International Journal of Geriatric Psychiatry. 15(S1): S9–S13.

Carillo, M.C., E. Dishman and T. Plowman. 2009. Everyday technologies for Alzheimer's disease care: Research findings, directions and challenges. Alzheimer's & Dementia. 5(6): 479–488.

Cavallo, M., E. Zanalda, H. Johnson, A. Bonansea and C. Angilletta. 2016. Cognitive training in a large group of patients affected by early-stage Alzheimer's disease can have long-lasting effects: A case-control study. Brain Impairment. 17(2): 182–192.

Chan, M., E. Campo, D. Estève and J.Y. Fourniols. 2009. Smart homes—current features and future perspectives. Maturitas. 64(2): 90–97.

Chan, M., D. Estève, J.Y. Fourniols, C. Escriba and E. Campo. 2012. Smart wearable systems: Current status and future challenges. Artificial Intelligence in Medicine. 56(3): 137–156.

Cook, D.J. and S.K. Das. 2007. How smart are our environments? An updated look at the state of the art. Pervasive and Mobile Computing. 3(2): 53–73.

Côté, L. and P. Hottin. 2012. Guide pour les proches aidants et les intervenants : Problèmes rencontrés dans la maladie d'Alzheimer. Le Centre de santé et de services sociaux, Institut universitaire de gériatrie de Sherbrooke. 32(1).

Coughlin, J., L.A. D'ambrosio, B. Reimer and M.R. Pratt. 2007. Older adult perceptions of smart home technologies: implications for research, policy & market innovations in healthcare. Conf. Proc. IEEE Eng. Med. Biol. Soc. 2007: 1810–5.

Demiris, G., N. Charness, E. Krupinski, D. Ben-Arieh, K. Washington, J. Wu and B. Farberow. 2010. The role of human factors in telehealth. Telemedicine and E-Health. 16(4): 446–453.

DeVaul, R.W., A. Pentland and V.R. Corey. 2003. The Memory Glasses: Subliminal vs. Overt Memory Support with Imperfect Information. International Symposium of Wearable Computers (ISWC'03). p. 146.

Doherty, A.R., S.E. Hodges, A.C. King, A.F. Smeaton, E. Berry, C.J. Moulin, A. Lindley, P. Kelly and C. Foster. 2013. Wearable cameras in health: the state of the art and future possibilities. American Journal of Preventive Medicine. 44(3): 320–323. Elsevier.

Donnelly, M., T. Magherini, C. Nugent, F. Cruciana and C. Paggeti. 2011. Annotating sensor data to identify activities of daily living. pp. 41–48. *In*: Abdulrazak, B., S. Giroux, B. Bouchard, H. Pigot, M. Mokhtari (Eds.). Toward Useful Services for Elderly and People with Disabilities. Springer-Verlag Berlin, Heidelberg.

Evans, J.S.B. 2003. In two minds: dual-process accounts of reasoning. Trends in Cognitive Sciences. 7(10): 454–459.

Frisardi, V. and B.P. Imbimbo. 2011. Gerontechnology for demented patients: Smart homes for smart aging. Journal of Alzheimer Disease. 23(1): 143–146.

Gentry, T. 2009. Smart homes for people with neurological disability: State of the art. NeuroRehabilitation. 25(3): 209–217.

Gitlin, L.N., L. Winter, M.P. Dennis, M. Corcoran, S. Schinfeld and W.W. Hauck. 2002. Strategies used by families to simplify tasks for individuals with Alzheimer's disease and related disorders psychometric analysis of the task management strategy index (TMSI). The Gerontologist. 42(1): 61–69.

Haigh, K.Z., L.M. Kiff, J. Myers, V. Guralnik, K. Krichbaum, J. Phelps, T. Plocher and D. Toms. 2004. The independent lifestyle assistant: lessons learned. Assist. Technol. 18(1): 87–106.

Hoey, J., P. Poupart, A.V. Bertoldi, T. Craig, C. Boutilier and A. Mihailidis. 2010. Automated handwashing assistance for persons with dementia using video and a partially observable markov decision process. Computer Vision and Image Understanding (CVIU), Elsevier. 114(5).

Holthe, T. and K. Engedal. 2004. ENABLE report: Enabling technologies for people with dementia. National Report on Results from Norway.

Huang, Y.C., C.-H. Lu, T.-H. Yang, L.-C. Fu and C.-Y. Wang. 2010. Context-aware personal diet suggestion system. pp. 76–84. *In*: Lee, Yeunsook., Bien Zenn, Z., M. Mokhtari, J.T. Kim, M. Park, H. Lee and I. Khalil (Eds.). Aging Friendly Technology for Health and Independence. Springer-Verlag Berlin Heidelberg.

Imbeault, H., H. Pigot, N. Bier, L. Gagnon, N. Marcotte, S. Giroux and T. Fülöp. 2011. Interdisciplinary design of an electronic organizer for persons with Alzheimer's disease. pp. 137–144. *In*: Abdulrazak, B., S. Giroux, B. Bouchard, H. Pigot and M. Mokhtari (Eds.). Toward Useful Services for Elderly and People with Disabilities. Springer-Verlag Berlin Heidelberg.

James, A.B. 2008. Activities of daily living and instrumental activities of daily living. pp. 538–578. *In*: Crepeau, E.B., E.S. Cohn, B.B. Schell (Eds.). Willard and Spackman's Occupational Therapy. Philadelphia: Lippincott, Williams and Wilkins.

Joseph, P. and F. Monk. 2008. Technological opportunities for supporting people with dementia who are living at home. International Journal of Human-Computer Studies. 66(8): 571–586. Elsevier.

Kidd, C.D., R. Orr, G.D. Abowd, C.G. Atkeson, I.A. Essa, B. MacIntyre, E. Mynatt, T.E. Starner and W. Newstetter. 1999. The Aware Home: A living laboratory for ubiquitous computing research. Cooperative buildings. Integrating Information, Organizations and Architecture, Lecture Notes in Computer Science. Springer Berlin Heidelberg. 1670: 191–198.

Lozano, H., I. Hemaez, J. Camarena, I. Diez and E. Nava. 2012. Identification of sounds and sensing technology for home-care applications. pp. 74–81. *In*: Bravo, J., R. Hervas and M. Rodriguez (Eds.). Ambient Assisted Living and Home Care. Springer-Verlag Berlin, Heidelberg.

Lopresti, F.E., A. Mihailidis and N. Kirsch. 2004. Assistive technology for cognitive rehabilitation: State of the art. Neuropsychological Rehabilitation. 14(1-2): 5–39.

Mihailidis, A., J.N. Boger, T. Craig and J. Hoey. 2008. The COACH prompting system to assist older adults with dementia through handwashing: An efficacy study. BMC Geriatrics. 8(28).

Morris, M., J. Lundell and E. Dishman. 2004. Catalyzing social interaction with ubiquitous computing: A needs assessment of elders coping with cognitive decline. pp. 1151–1154. *In*: Proceedings of the Paper Presented at CHI, Vienna, Austria.

Navarette, I., J.A. Rubio, J.A. Boita, J.T. Palma and F.J. Campuzano. 2012. Ambient modeling a risk detection system for elderly's home-care with a network of timed automata. pp.

82–89. *In*: Bravo, J., R. Hervas and M. Rodriguez (Eds.). Ambient Assisted Living and Home Care. Springer-Verlag Berlin, Heidelberg.

O'Neill, B. and A. Gillespie. 2008. Simulating naturalistic instruction: The case for a voice mediated interface for assistive technology for cognition. Journal of Assistive Technologies. 2(2): 22–31.

Pilotto, A., G. D'Onofrio, E. Benelli, A. Zanesco, A. Cabello, M.C. Margeli and D. Kilias. 2011. Information and communication technology systems to improve quality of life and safety of Alzheimer's disease patients: A multicenter international survey. J. Alzheimer's Dis. 23(1): 131–141.

Ready, E. Rebecca and R.O. Brian. 2003. Quality of life measures for dementia. Health Qual Life Outcomes. 1: 11. Published 2003 Apr 23. doi: 10.1186/1477-7525-1-11.

Rialle, V. 2007. Technologie et Alzheimer: Appréciation de la faisabilité de la mise en place de technologies innovantes pour assister les aidants familiaux et pallier les pathologies de type Alzheimer (Doctoral dissertation, Université René Descartes-Paris V).

Rialle, V., C. Ollivet, C. Guigui and C. Herve. 2008. What do family caregivers of Alzheimer's disease patients desire in smart home technologies? Contrasted results of a wide survey. Methods Inf. Med. 47(1): 63–69.

Rougier, C., E. Auvineti, J. Rousseau, M. Mignotte and J. Meunier. 2011. Fall detection from depth map video sequences. pp. 121–128. *In*: Abdulrazak, B., S. Giroux, B. Bouchard, H. Pigot and M. Mokhtari (Eds.). Toward Useful Services for Elderly and People with Disabilities. Springer-Verlag Berlin Heidelberg.

Scherer, M.J. 2005. Living in a state of stuck: How technology impacts the lives of people with disabilities. Cambridge, MA: Brookline.

Selmès, J. and C. Derouesné. 2004. Réflexions sur l'annonce du diagnostic de la maladie d'Alzheimer. Psychologie & NeuroPsychiatrie du vieillissement. 2(2): 133–140.

Tee, K., K. Moffatt, L. Findlater, E. MacGregor, J. McGrenere, B. Purves and S.S. Fels. 2005. A visual recipe book for persons with language impairments. Papers presented at the Conference on Human Factors in Computing Systems, held in Portland, Ore.; April 2–7.

Tinetti, M.E., D. Richman and L. Powell. 1990. Falls efficacy as a measure of fear of falling. Journal of Gerontology. 45(6): P239–P243.

Topo, P., K. Saarikalle, O. Mäki and S. Parviainen. 2004. ENABLE report: Enabling technologies for people with dementia. Report of Assessment Study in Finland.

Tran, Q., E. Mynatt and G. Calcaterra. 2007. Using memory aid to build memory independence. Human-Computer Interaction. Interaction Design and Usability, Lecture Notes in Computer Science, Springer; Berlin/Heidelberg. 4550: 959–96.

Tsui, K.M. and H.A. Yanco. 2010. Prompting devices: A survey of memory aids for task sequencing. In QoLT International Symposium: Intelligent Systems for Better Living.

Van Tassel, M., J. Bouchard, B. Bouchard and A. Bouzouane. 2011. Guidelines for Increasing Prompt Efficiency in Smart Homes According to the Resident's Profile and Task Characteristics. 6719: 112–120.

Walker, M.D., S.S. Salek and A.J. Bayer. 1998. A review of quality of life in Alzheimer's disease. Part 1: Issues in assessing disease impact. Pharmacoeconomics. 14(5): 499–530.

Wherton, J.P. and A.F. Monk. 2008. Technological opportunities for supporting people with dementia who are living at home. International Journal of Human-Computer Studies. 66(8): 571–586.

Zhou, F., H.-I. Yang, J.M.R. Alamo, J.S. Wong and C. Chang. 2010. Mobile Personal Health Care System for Patients with Diabetes. pp. 94–101. *In*: Lee, Yeunsook., Bien Zenn, Z., M. Mokhtari, J.T. Kim, M. Park, H. Lee and I. Khalil (Eds.). Aging Friendly Technology for Health and Independence. Springer-Verlag Berlin, Heidelberg.

3

Pervasive Computing and Ambient Physiological Monitoring Devices

Sung Jae Isaac Chang,[1,2,3,*] *Jennifer Boger*,[2,3] *Jianfeng Qiu*[4] and *Alex Mihailidis*[1,2,3]

1. Introduction

The Public Health Agency of Canada reports that chronic diseases are one of the most serious problems to attend in the health care system. In 2012, 85 per cent of the older adults aged between 65 and 79 years and 90 per cent of seniors aged over 80 years had at least some chronic disease (Public Health Agency of Canada 2014). Examples of chronic diseases include asthma, bronchitis, heart disease, diabetes, high blood pressure, post-stroke effects and Alzheimer's disease. Such illnesses are difficult and

[1] Institute of Biomaterials and Biomedical Engineering, University of Toronto. 12th floor 550 University Avenue Toronto, Ontario, Canada. Email: Isaac.chang@mail.utoronto.ca

[2] Intelligent Assistive Technology and Systems Lab, Dept. of Occupational Science & Occupational Therapy, University of Toronto. 160-500 University Avenue Toronto, Ontario, Canada. Email: Jen.boger@utoronto.ca

[3] Toronto Rehabilitation Institute, Toronto, 190 Elizabeth ST, Ontario, Canada, Email: Alex.Mihailidis@utoronto.ca

[4] Centre for Global eHealth Innovation, University Health Network. 160-500 University Avenue Toronto, Ontario, Canada, Email: hqiu@ehealthinnovation.org

* Corresponding author

expensive to manage, both for the patient and for the healthcare system. Additionally, in industrialized countries, such as Canada, the US, West European countries including the UK, and Japan, people who are 60-years old or older are expected to account for between 26.8 per cent and 44.0 per cent of the population by 2050, as compared to 17.8 per cent and 26.4 per cent in 2005 (Mercado et al. 2007). The greater prevalence in chronic health problems experienced by older adults is predicted to cause a significant rise in already expensive health care costs and overextended systems.

A study conducted in the US showed that more than 90 per cent of adults aged 65 or more prefer to live in their own house with their family (Farber et al. 2011). Moreover, it has been demonstrated that so long as conditions are properly managed, aging-in-place results in better health outcomes than long-term care placements (Marek and Rantz 2000). Supporting the monitoring and management of chronic health conditions enables aging-in-place, where older adults live healthily and independently in the place of their preference while simultaneously reducing the healthcare costs on long-term care and hospitals.

To address this need, there is a trend towards developing technologies that support well-being and independent living, such as remote monitoring of vital signals. A goal of remote monitoring is to identify potential problems as close to their onset as possible, thereby enabling early interventions that can improve outcomes for both the person and the healthcare system. Remote monitoring also facilitates ongoing data collection in a stable condition. This permits long-term and continuous understanding of a person's health, which could lead to better disease management (both by the clinicians and through self-management) and potentially fewer visits to doctors and hospitals.

This chapter presents zero-effort technologies (ZETs) for performing ambient physiological monitoring by discussing how two conditions commonly associated with aging, namely dementia and cardiovascular disease, could benefit from this type of monitoring before introducing the concepts of remote physiological monitoring, pervasive computing and ZETs. This is followed by discussion on three form factors used in ambient physiological monitoring—a standing-form factor (i.e. weight scale and floor tile), bed and sitting-form factor (i.e. chair, couch and toilet seat). Subsequently activity monitoring and the general clinical relevance of zero-effort physiological monitoring are discussed before the chapter concludes with future directions for research.

1.1 Chronic Conditions and Remote Monitoring

While ambient physiological monitoring can be used in the management of several medical conditions, two conditions of special concern are dementia and cardiovascular disease. People with dementia can benefit tremendously from the effortless way the ambient monitoring technology autonomously samples and transmits physiological parameters to stakeholders of interest, regardless of the person's cognitive abilities. Cardiovascular disease symptoms can fluctuate and thus can benefit from continuous collection of the physiological parameters, such as heart rate and blood pressure to quickly make important clinical diagnosis and assessment of the disease.

1.1.1 Dementia

Dementia is a general term that refers to a condition where a person's mental abilities and thinking processes decline to a point as to affect his or her daily functions (Alzheimer's Association 2014). The number of people with dementia was estimated at 44 million globally in 2014 and the cost to support dementia in the US alone was recorded at $604 billion in 2010 (Alzheimer's Disease International 2014). This number is projected to increase to 132 million, more than for the people with dementia by 2050 (Alzheimer's Disease International 2014). Dementia and its management have significant social and economic impacts, ranging from healthcare to personal levels.

Symptoms of dementia usually manifest first as memory loss, followed by declining cognitive processes, e.g. judgement, decision making, personality and sleep patterns (Alzheimer's Association 2014). However, the severity and rate of change of these symptoms vary widely between individuals as well as from day to day. Loss of memory, planning, activity initiation and other abilities make it difficult or impossible for people with dementia to remember to perform the activities and tasks required of them to maintain their well-being. As symptoms worsen, the burden of managing dementia shifts from a person with dementia to caregivers. In particular, informal caregivers (i.e. family members or friends) assume the tasks required to keep a person with dementia safe and healthy; however, caregiving for a loved one can result in significant mental and physical stress. It is reported that 15 to 32 per cent of the caregivers show signs of clinical depression and 40 to 75 per cent suffer from some form of a mental illness (Alzheimer's Disease International 2009). A study based on more than 1,500 caregivers showed that the caregivers who were caring for people with dementia had more negative associations with their personal time, employment issues and family conflicts than the caregivers who took care of people with other conditions (Ory et al. 1999). Therefore, when considering how to support dementia, it is important to find solutions that complement caregivers as well as people with dementia; namely, interventions need to support the abilities of people with dementia without increasing the burden on caregivers. One of the ways to tackle this issue is to use assistive devices at home that are automated and unobtrusive for the users, ensuring that the operation of the device does not burden the caregiver but assists the person with dementia.

1.1.2 Cardiovascular disease

Cardiovascular disease is the leading cause of morbidity and mortality globally across all age-groups (Heart and Stroke Foundation of Canada 2013). Within Canada, the disease costs more than $20.9 billion per year (based on 2010 statistics), accounting for both direct (e.g. hospital cost, clinician services) and indirect (e.g. lost wages, reduced productivity) impacts (Heart and Stroke Foundation of Canada 2015). Nearly 23 per cent of Canadian seniors reported having some type of heart disease in 2009 and more than 50 per cent reported to have two key risk factors—high blood pressure and sedentariness (Public Health Agency of Canada 2014). In the U.S., it is estimated that 5.7 million people have a related condition and the disease claims 300,000 lives each year; this translates into the third leading cause of death in the US (Kung et al. 2008; Lloyd-Jones et al. 2009).

One of the most prominent cardiovascular diseases is heart failure (HF). HF is defined as a progressive condition where the heart is not able to supply sufficient blood to the tissues (Heart and Stroke Foundation of Canada 2013). This can cause lack of oxygen in various organs, including the brain and muscles, resulting in dizziness and weakness of the body, as well as shortness of breath due to improper circulation of blood in the lungs (Heart and Stroke Foundation of Canada 2013). The body compensates in several ways in an effort to get enough blood to its tissues, including fluid retention caused by increased vasopressin in the blood and activation of the sympathetic nervous system, leading to higher contractility of the heart and higher cardiac output. HF can lead to kidney and liver damage or failure, heart valve and rhythm problems and a host of secondary problems. There were 34.8 million new cases of HF reported in the US in 2000 and this is projected to increase to 77.2 million in 2040 (Owan and Redfield 2005). In the US alone, the healthcare cost for HF was estimated to be $37 billion in 2009, which is higher than the cost associated with any other disease (Lloyd-Jones et al. 2009).

HF disease management programs, including home-based monitoring, have been reported to be effective in improving self-management, quality of life and health outcomes (Domingo et al. 2011). For example, a person with HF is usually assessed for excess fluid volume at the clinic by inserting a catheter with a pressure sensor (Remme and Swedberg 2001). However, since it is necessary to monitor the condition more frequently, people with HF are advised to measure their body weight daily at home. Fluctuations in body weight often signal changes in the condition; a person with cardiac disease typically gains 2–5 pounds one or two weeks prior to his hospitalization (Chaudhry et al. 2007). Goldberg et al. (2003) performed a study involving 280 people with severe HF and showed that daily body weight measurement at home effectively reduced mortality rate by 56.2 per cent. Other parameters could benefit from ongoing daily monitoring, such as heart rate, respiration rate, blood pressure and temperature. Monitoring these parameters could help to avoid acute decompensation—a state where the heart fails to maintain regular blood circulation (University of Ottawa Heart Institute 2015) as well as to inform whether the condition is being managed appropriately.

1.1.3 Remote monitoring using zero-effort technologies

Dementia and HF are just two examples of chronic conditions that could benefit from remote monitoring. In response, there have been major developments in mobile health monitoring and ambient home health monitoring. Both these fields have the common themes of detecting the early onset of a disease and/or managing an existing condition.

Mobile health monitoring refers to collecting and assessing one's health information using mobile devices, such as a smart phone or other wearable sensors. Recent examples are the Apple Watch from Apple and the Gear S watch from Samsung, both of which can record and process a person's vital signs. The merits of these technologies is that they provide features that allow a user to personalize and customize the device to fit to his or her needs. For instance, Samsung's Gear S and the Apple Watch are equipped with a heart rate sensor, global positioning system (GPS), microphone, accelerometer and other capabilities that can provide various measurements. Combined with a smart phone, it is possible to personalize monitoring to reflect personal interests and goals (Apple 2015; Samsung 2015). When devices such as these are incorporated into the monitoring of chronic diseases, the level of personalization and mobility could allow both clinicians and individuals to gain better insight into the health of the person with the disease. Despite the potential benefits of mobile health monitoring, it is important to acknowledge that the fundamental presumption underneath the designs of these devices is that the user is physically and cognitively capable of operating them. Some issues identified with mobile and wearable monitoring devices are forgetting to wear the device, missing the prescribed measurement schedule and incorrect device operation (Alwan 2009). These issues are likely to be exacerbated for older adults, who are generally less comfortable with operating new technologies and have a greater prevalence of cognitive disabilities, such as dementia, that would make using such a device difficult or impossible.

Ambient home health (physiological) monitoring is an emerging area of development that refers to monitoring one's health by embedding sensors into objects that a person normally and naturally interacts with, such as furniture, house appliances and the home environment itself. As these sensors and devices are unobtrusively incorporated throughout a person's environment. Ambient home physiological monitoring is considered to be a type of pervasive computing and is also known as ubiquitous computing or 'everyware' (Greenfield 2010). Pervasive computing is a type of informatics where sensing, computing and interfacing are not restricted to one particular device or location but spread throughout the environment. The benefit of this format is that the person being monitored is not restricted to a specific medical device for health monitoring but can achieve monitoring via interaction with everyday objects, like a chair or bed. Having sensors around the home also provides temporal regularity of measurement; for example, the person being monitored will generally interact with objects, such as a bed, in a regular manner each day (e.g. going to sleep each night). This can provide regular and contextual data that enables clinicians to have a more comprehensive, ongoing idea of a person's health.

In addition to being a type of pervasive computing, most ambient home physiological monitoring devices are being developed as zero-effort technologies (ZETs) (Mihailidis et al. 2011). A ZET is defined as technology that frees the user completely or almost completely from the effort needed to operate the technology while achieving the task he or she would like to do. For example, the current conventional way to measure blood pressure involves explicit interaction with a blood pressure machine by wrapping an air cuff around the arm and pressing a button on the machine, which displays blood pressure values a few seconds later. Thus the person needs to go to the location where the blood pressure machine is, wrap the cuff around the arm, press button, make a note of the blood pressure and log this result so his or her clinician can see it. All of these steps can be considered to be an 'effort' that the person must put in to obtain his or her blood pressure. In contrast, a ZET is able to autonomously, automatically and effortlessly obtain the person's blood pressure without the person having to change anything about his daily routine. One of the devices introduced in this chapter is a floor tile that can measure blood pressure when a person stands on it. By positioning the tile in front of the bathroom or kitchen sink, the floor tile measures the heart rate and blood pressure each time the person washes his hands, brushes his teeth, or washes the dishes. The person being monitored does not put in any effort to obtain the blood pressure readings, which can be sampled several times a day at different times of the day. In this way, ZETs are able to obtain more comprehensive physiological data without adding any burden to the person being monitored or (if applicable) his caregiver. By definition, ZETs do not include wearable devices, which require effort from the user (e.g. wearing the device, remembering to wear the device, changing or charging batteries, etc.). As this chapter focuses on ZETs, topics like wearable devices or portable-form factors are not discussed.

Ambient physiological monitoring represents valuable and versatile tools for supporting the monitoring and management of chronic health conditions. In terms of dealing with dementia, zero-effort ambient physiological monitoring can free the person with dementia and his or her caregivers from the burden of learning and remembering how to operate medical devices, as well as the burden of keeping track of the medical measurement schedules. The effort that is saved by the ambient physiological monitoring can be used to address other important matters of living, thus effectively increasing the quality of life for the person with dementia as well as the caregivers. The same logic can be applied to people with HF. Zero-effort ambient monitoring can monitor the user's vital signs effortlessly and regularly; the mental and physical energy and time saved can then be allocated to other activities. Most importantly, it is likely that continuous, ambient physiological monitoring will enable early detection of an onset or worsening of a chronic disease, helping to design interventions that take place in a timely manner.

1.2 Physiological Signals used in Non-Vision Related Ambient Physiological Monitoring

Conventional methods of assessing health use standard medical equipment to measure physiological signals, such as an electrocardiogram (ECG) to measure heart rate or a photoplethysmogram (PPG) to measure oxygen saturation level. More recent advancement of health assessment uses wearable sensors to make measurements. However, wearable sensors may not be ideal for monitoring older adults, especially those with cognitive impairments, because of the issues mentioned in Section 1.1.3. As discussed previously, ZETs are a promising new option for monitoring chronic conditions as they are embedded into the environment, measure signals unobtrusively and require little or no effort on the part of the user and these devices are possibly more suitable for collecting health information from older adults.

Any monitoring device requires acquisition of physiological signals in order to extract the relevant physiological parameters (e.g. heart rate). Despite the numerous possible form factors, which are discussed in Sections 2, 3 and 4, there are a few basic, conventional physiological signals that are collected, including:

- electrocardiogram (ECG)
- ballistocardiogram (BCG)
- photoplethysmogram (PPG)
- respiration signal

Each of these signals is introduced before discussing how these signals are acquired through different devices.

1.2.1 Electrocardiogram (ECG)

ECG is a recording of the electrical activity of the heart. As the electrical currents travel around the heart to contract the heart muscle cells, the currents' electromagnetic field is propagated throughout the body. The propagation of the electromagnetic field reaches the skin, which can be detected by attaching at least two electrodes on the skin. Attaching two electrodes to record ECG can be explained as viewing a one-dimensional slice of a three-dimensional electromagnetic contours. This means by shifting the location of the electrodes, it enables one to examine the cardiac activity from a different perspective.

One can record multiple ECG signals by attaching multiple electrodes on the body where each signal reveals information about the heart activity. One of the well-known configurations of measuring ECG is 3-lead ECG. Because there are three leads or electrodes, three different ECG signals are recorded. A 3-lead ECG is commonly used because it is relatively simple to set up and signals provide different perspectives of the heart activity, enabling the clinician to assess the heart's function. A more sophisticated method of measuring ECG is 12-lead ECG where 12 electrodes are placed on the defined locations of the torso, allowing the observer to have much more in-depth information on how the heart is functioning at various stages of its cycle.

1.2.2 Ballistocardiogram (BCG)

Ballistocardiogram (BCG) is defined as mechanical vibration of the body due to the reaction force of the heart pumping blood (Starr et al. 1939). At each heartbeat, the chambers (i.e. ventricles) of the heart containing blood contracts, push out blood into the vascular system. As this event takes place, the body experiences a minute mechanical reaction force and the recoding of this signal is called BCG. BCG can be used to find the heart rate and to reveal information on the cardiac output (Inan et al. 2009). It is estimated that BCG is approximately 0.24 N of oscillation force while standing (2008). BCG can be detected while the person is lying down, sitting, standing or by wearing a BCG measuring device. BCG is actually a three-dimensional signal and the BCG recording will capture one of the three components or a mixture of the components, depending on the body orientation at the time of measurement. The longitudinal (i.e. up and down) BCG is captured when it is recorded while standing and a combination of anterior-posterior (i.e. front and back) and lateral (i.e. side to side) BCG is captured from the lying down position. To avoid confusion, the different components of BCG will be called BCG throughout the chapter.

Since BCG is a force-based signal, it can be detected even if there is a medium between the skin and the sensor (e.g. clothing) since the force is propagated through the medium. The point where the signal propagates depends on the characteristics of the medium.

Although measuring BCG provides an advantage of non-contact to the skin, it has a drawback of being susceptible to movement artifacts. This is because having an approximate amplitude of about 0.24 N, even a small force can easily overpower the signal, causing a high distortion. Algorithms, that have been developed to address this problem, will be discussed in Sections 2.1 and 3.1.

1.2.3 Photoplethysmogram (PPG)

Photoplethysmogram (PPG) is a physiological signal that reflects the oscillation of the blood volume in the vascular system (i.e. blood vessels) due to heart activity. During the systolic phase of the cardiac cycle, which is immediately after the ventricles contract to send blood throughout the body, the blood is pushed through the arteries and the amount of blood in the vascular system increases. Following the systole, the blood fills the heart chambers and the heart readies for the next contraction. In this phase, the amount of the blood present in the vascular system becomes relatively less. It is difficult and requires invasive methods to measure this change of blood volume in the arteries. The change of the blood volume in the capillary beds, on the contrary, is reflected to the skin and can be detected using appropriate tools. This detection of blood-volume change in capillary beds using a non-invasive sensor is called PPG. PPG provides the heart rate as well as the oxygen saturation level, which is defined as how much oxygen is being carried by the blood at the time of measurement.

In order to transform the micro-fluctuation of the blood volume into an electrical signal, PPG uses a light source and a light detecting sensor. A light source, such as a light emitting diode (LED), is placed on the skin and shines light into the body. A light-detecting sensor, such as a photodiode placed beside or on the opposite side of the light source, then detects the light transmitted through or bounced back from the skin. The intensity of the transmitted light changes according to the change in the blood volume and is captured by the sensor. The micro-fluctuation of blood volume is now converted into an electrical signal that can be further processed and recorded. The caveat when collecting PPG is that ambient light sources must be completely blocked to allow micro-fluctuation of the light intensity to be captured properly.

1.2.4 Respiration signal

In the context of ambient physiological monitoring, respiration refers to the breathing of a person. Respiration signal is a waveform that reflects the expansion and relaxation of the lungs while a person breathes. It provides insights into the respiratory functions of the person and enables detection of any abnormal breathing pattern. The signal can be especially useful in determining if the person has any breathing problem while sleeping, such as obstructive sleep apnoea.

Respiration signal is measured either by detecting the air flow though the mouth or by detecting the expansion of the chest as a person breathes. Devices, such as a spirometer, can be used to measure the air flow through the mouth while breathing. The chest expansion during breathing can be detected by wearing a chest strap with a sensor and measuring how much the chest expanded while breathing. In a recent development, respiration signal was found in other physiological signals, such as BCG and ECG, and ways to extract the respiration signal from these physiological signals have been investigated (refer to Section 3.1).

2. Standing Form Factors

People often stand in various spots in their home to complete a variety of ADL, such as hand washing, tooth brushing and kitchen tasks. Two form factors being developed to perform ambient monitoring while a person is standing are the body weight scale and a floor tile. The rationale behind using a weight scale is its familiarity and usability; moreover, many health conditions require people to measure their weight daily. In addition to weight, there are a number of other physiological parameters that could be useful to measure, such as heart rate and blood pressure. In addition to the weight, the main physiological signals collected from these devices are BCG and ECG. BCG alone gives heart rate and a combination of BCG and ECG provides an estimate of the systolic blood pressure (SBP). This section will discuss how BCG can be collected from a weight scale or floor tile, how BCG and ECG can be combined to give estimation of SBP and some of the challenges of measuring signals in the standing posture.

2.1 BCG Acquisition

BCG is the minute force generated by the heart during the cardiac cycle with the heart contracting and expanding to move blood through the body. A transducer that can convert this force to electrical energy is used to collect the signal. While different force transducers are available, thus far load cells have been used in weight scale and floor tile applications. In most of the off-the-shelf bathroom weight scales there are one or more load cells installed on the bottom of the scale, often one at each corner. As a person stands on the scale, the load cells and an analog circuit convert the weight into an electrical signal that can be measured and shown on an LED screen as a value that the user can comprehend, such as kilograms or pounds. The same concept is used to collect BCG.

BCG can be detected as a very small oscillation added to a person's weight. To illustrate, suppose an object with weight similar to a typical human body rests on a scale. BCG force has an approximate amplitude of 0.24 N or 24 grams, which is equivalent to lightly tapping on the object with a finger. When a person stands still on a weight scale, no change in the displayed weight is observed because the scale is not sensitive enough to detect it; however, in reality, the weight on the scale oscillates a very small amount because of BCG. Modifying the load cells and circuit to increase the sensitivity of the scale enables the capture of BCG. BCG collected from a weight scale or floor tile can be used to measure the heart rate by looking at the periodic oscillation of the BCG signal. A typical BCG waveform is shown in Fig. 1, along with the ECG for comparison.

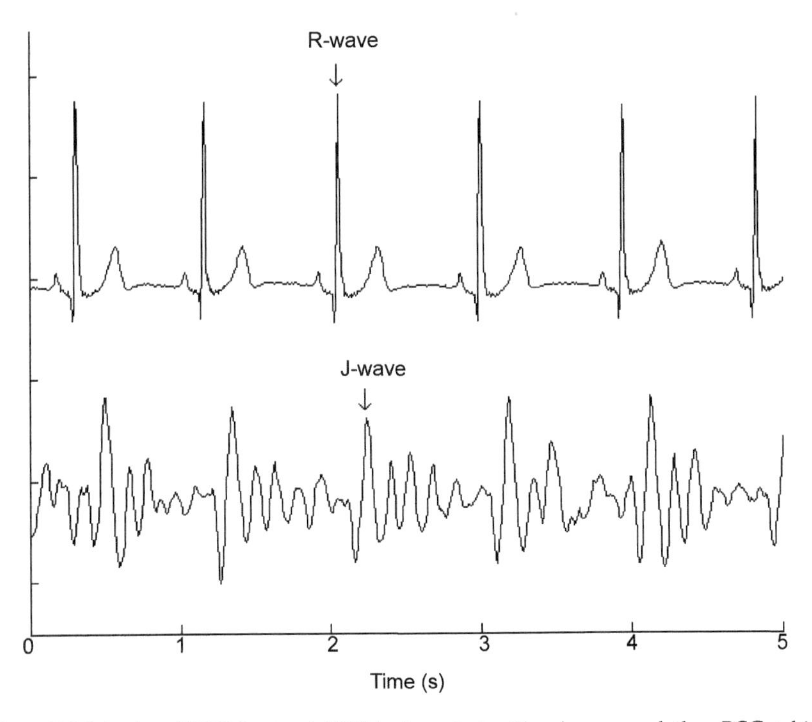

Fig. 1. ECG (top) and BCG (bottom): ECG is characterized by sharper peak than BCG, which has smoother waveform than ECG; the R-wave of ECG, which indicates the beginning of the ventricle muscle contraction of the heart, precedes the J-wave of BCG which is the result of strongest reaction force to the ejection of the blood from the heart.

Several researchers have developed floor tile or weight scales that are able to detect BCG. For example, Inan et al. (2009) developed a circuit involving four stages of signal processing, which included an instrumentation amplifier, AC coupling via active integrator-based feedback loop, a Sallen-and-Key low pass filter and a non-inverting amplifier. The total gain of the system is 11000 and the bandwidth is between 0.1 to 25 Hz (Inan et al. 2009). Gonzalez-Landaeta et al. (2008) designed another circuit for detecting BCG which involves AC-coupled differential amplifier, first-order low pass filter, and a non-inverting amplifier. The total gain is 7500 and the bandwidth is between 0.1 to 10 Hz (2008). The main difference between these designs is that the former had a higher gain and a higher upper cut-off frequency (i.e. 25 Hz), resulting in a slightly larger BCG signal and more of a higher frequency information. As both designs have digital processing following the analog circuit, the moderate difference in gain is less important. Even though the upper cut-off frequencies were different, both designs were successful in capturing a reliable BCG signal, meaning, the majority of the frequency component of BCG reside under 10 Hz and the information above 10 Hz did not contain any significant value. In other words, having signal components below 10 Hz means that BCG is composed of smooth waveform without sharp peaks, which is unlike those observed in ECG. Without sharp peaks, BCG is more susceptible to movement artifacts compared to ECG, where the frequency component of a movement artifact signal is often less than 10 Hz (i.e. a person can move his or her arm only a few times in one second). The smooth curvature of BCG also makes finding the exact location of heart activity more difficult than it is for ECG, which means that heart rate values obtained from BCG have higher variance compared to those obtained from ECG. This drawback can be circumvented by averaging beat-to-beat heart rate values. Thus BCG can be a viable method to measure a person's heart rate, though, a more precise method is to use ECG.

2.2 R-J Interval Method: Estimation of SBP using BCG and ECG

In addition to the heart rate, Shin et al. (2009) recently developed a physiological monitoring method that enables SBP to be estimated by combining ECG and BCG data. In ECG, the largest peak is called the R-wave while the equivalent wave in BCG is called the J-wave (Fig. 1). The time between the R-wave of ECG and J-wave of BCG is defined as R-J interval (Inan et al. 2008). A number of findings have been made on the R-J interval and its physiological relevance to other parameters. In a study involving 15 healthy adults, it was found that the R-J interval is strongly correlated with pre-ejection period (PEP), which is the time between the

ECG R-wave and ejection of blood out of the heart (Inan et al. 2008). Etemadi et al. (2009) showed that R-J interval was correlated to PEP with r^2 value of 0.75 based on 4 subjects. Shin et al. (2009) found that the R-J interval is inversely proportional to SBP and used this information to estimate SBP based on the R-J interval. The group performed an ordinary linear regression on R-J interval and SBP data for each subject and used the R-J interval values as inputs to predict the corresponding values of SBP. Based on 10 healthy adults, the estimation achieved the mean relative error of 8.4 per cent. Using the same method, the authors achieved 7.4 per cent error with standard deviation of 2.74 per cent based on 52 healthy adults. Even though it was possible to estimate the SBP based on R-J interval, the accuracy and reliability of the method need further verification to be used for clinical practice. For example, the repeatability of the relationship between R-J interval and SBP from the same subject is yet to be verified.

While it has been shown that there is a correlation between R-J interval and SBP, there is evidence that the relationship between the R-J interval and SBP is not proportional in every subject. Casanella et al. (2012) showed that for two subjects, only one subject showed correlation between the R-J interval to SBP whereas the other subject did not. Inan et al. (2008), Shin et al. (2009) and the authors have calculated the correlation between R-J interval and SBP using Valsalva manoeuvre (i.e. holding one's breath while applying pressure to the chest for a few seconds to cause a change in blood pressure) while Casanella et al. (2012) used paced respiration (i.e. breathing at a faster rate than when at rest) to change the blood pressure. The Valsalva manoeuvre creates changes in blood pressure over a period of tens of seconds whereas paced respiration changes the blood pressure more rapidly, within seconds. The study by Casanella et al. (2012) investigated if the relationship between R-J interval and SBP based on Valsalva manoeuvre was reflected in more rapid and smaller change induced by paced respiration. Based on the results, the relationships were not same. In the same article, the group also decomposed R-J interval into two smaller components to provide complementary information about the R-J interval. The complementary results also did not agree with the relationship recorded by Inan et al. (2008) and Shin et al. (2009). Further investigation into these issues is required since the sample sizes for these studies was small and in controlled conditions.

2.3 Challenges of Using the Weight Scale and Floor Tile

So far the use of conventional ECG to estimate SBP by using the R-J interval method has been discussed. However, for this method to be practically useful for ambient physiological monitoring devices, such as the weight scale or floor tile, two important obstacles must be overcome: (1) BCG signal is lost whenever there is even a small movement, such as turning one's head or moving one's fingers, and (2) electromyogram (EMG) hinders the collection of ECG when it is measured from the feet of the person standing on the scale or tile. Works investigating the problem of reducing or removing movement artifacts are discussed first.

2.3.1 Movement artifact removal from BCG

Efforts have been made to improve BCG signal quality and to identify when a movement artifact is present (and thus distorting BCG). Examples of techniques for processing BCG to improve signal quality include:

- Continuous wavelet transform (CWT) to extract heart rate and respiration rate from the standing BCG (Gilaberte et al. 2010)
- Seismic sensor to subtract vibrations from floor from the BCG signal, obtaining higher signal to noise ratio (SNR) (Inan et al. 2010)
- Use of EMG to flag where there is strong movement artifact in BCG (Inan, Kovacs and Giovangrandi 2010)
- Four additional strain gauges attached on the opposite side of the original load cells to detect the body sway in the anterior-posterior direction, which is subtracted from the BCG, removing the background drift (Wiard et al. 2011)

In Gilaberte et al.'s work (2010), CWT was used to identify the heart rate and respiration rate from BCG. This was achieved by decomposing BCG using CWT based on Daubechies 10 wavelet into a range of different scales. Daubechies 10 based CWT uses a specific waveform shape, also known as a wavelet, to extract frequency information from the waveform to analyze. This technique allows one to locate the frequency component for analysis within the waveform. The heart rate and respiration rate were identified by matching the scales to the gold standards. This technique is useful in providing information about the recorded BCG but does not directly tackle the movement artifact problem. Inan et al. (2010a) measured the vibration of the floor to reduce the noise embedded in BCG, using an adaptive filter. The study applied the filter while a subject stood on the weight scale while riding a bus. When the bus was mobile, the artifact was too severe and BCG was not detected. However, when the bus was parked with its engine on, the adaptive cancellation of the noise was able to remove vibration noise of the bus from the BCG. The same group used EMG signals from the lower body to automatically detect the motion artifact. Based on 14 subjects' BCG recordings, it was found that the root mean square (RMS) power of the lower-body EMG can be used as a reference to indicate how much movement artifact is contained in a BCG signal (2010). Another movement artifact identification algorithm was developed by the same group where body sway, especially in the anterior-posterior direction, of the subject can indicate the presence of motion artifact (Wiard et al. 2011). By removing the signals that contained this movement artifact, the researchers were able to increase the signal to noise ratio by 14dB.

In general, movement artifact removal techniques either reduce the noise embedded in a BCG signal or crop portions of the signal that are too noisy. Currently, there is a threshold level of noise that can be removed from the BCG; once the artifact increases beyond the threshold, the affected portion of the signal must be removed. Incorporation of new sensors and more advanced adaptive algorithms may achieve such improvements. Moreover, techniques developed specifically to tackle problems found in one form factor may well be transferable to another. For example, the techniques discussed above could be applied to bed form factors and visa-versa, as touched upon in Section 3.1.

2.3.2 Removal of EMG Noise from F-ECG

ECG is typically collected by attaching electrodes to the chest, which is the closest location to the heart. This configuration provides the best signal quality and very little noise is added to ECG. The further an electrode is moved away from the heart, the weaker the ECG becomes. The farthest location to detect ECG is the feet, where the signal is weak but is still discernible in a noise-free setting. However, a noise-free setting is a difficult scenario to achieve and feet-sampled ECG (F-ECG) is typically easily corrupted due to its relatively weak signal quality and higher susceptibility to motion-generated artifacts.

In order to obtain ECG from a weight scale, Shin et al. (2009) attached electrodes to the scale to measure F-ECG and found F-ECG was severely corrupted by the EMG created by foot muscles. The corrupted F-ECG was not usable for heart rate extraction as the R-wave (i.e. the largest component in ECG) was covered by the presence of the EMG noise. The authors have also researched obtaining F-ECG by attaching stainless steel electrodes to a floor tile. Although the collection of F-ECG was successful while the user is sitting and putting his or her feet on the tile, the ECG signal is corrupted by EMG when the user is standing. Not all prototypes collect ECG from the feet. Inan et al. (2011) installed hand rails to the scale via a wire and measured ECG from the rails. The hand rails provided better quality of ECG compared to that measured from the feet due to closer proximity of the hands to the heart than feet. However, in cases where the design is aimed to achieve a high degree of unobtrusiveness, such as the floor tile, the most viable point to retrieve ECG is through the feet; therefore EMG noise must be removed for the device to achieve its intended purpose.

Research has been done to remove EMG noise by taking an ensemble average of the ECG signal (Shin et al. 2009; Gomez-Clapers et al. 2012). Ensemble average is a technique to find the average of all cycles found in a periodic signal. The original periodic signal is first divided into multiple segments, each of which contains a single cycle. Segments are added together then divided by the number of cycles. When used with a large enough period, the technique removes white noise (i.e. the noise that has an average of zero) from the signal. Because EMG is similar to white noise, this technique can be used to remove EMG noise from the signal to obtain a clean average ECG. The technique of ensemble average has been used to clean physiological signals, such as PPG (Sola et al. 2009). In other works, the authors have investigated how to remove the EMG noise from F-ECG using wavelet filter (i.e. a digital noise removal technique). The wavelets used were Morlet wavelet and an individual-specific ECG wavelet based on the ensemble average of ECG from the chest. Preliminary results showed signal improvements for 8 out of 14 subjects. This method is promising, however, it can only remove noise when the R-wave is still visible even in the presence of noise. If the R-wave is hidden by noise, which is the issue that the authors of this chapter are trying to address, then the method is not effective.

2.4 Current Research

Several groups have developed weight scale monitoring devices and evaluated them with healthy adults. Both Inan et al. (2009) and Gonzalez-Landaeta et al. (2008) developed weight scales that can measure weight and heart rate while Shin et al. (2009) developed a weight scale that can measure weight, heart rate and estimate blood pressure. Gonzalez-Landaeta et al. (2008) compared BCG from 20 subjects to their ECG; the heart beat interval (i.e. reciprocal value of heart rate) from BCG based on 17 subjects had a mean error of 0.1 ms with 95 per cent interval of ± 21 ms. This is equivalent to 0.006 beat per min error with standard deviation of ± 1.26 beats per min, which means that based on a typical 70 beats per minute heart rate, the error is less than 0.01 per cent with standard deviation of 1.8 per cent. As discussed in Section 2.2, SBP estimation was able to approximate SBP with 8.4 per cent error. It is interesting that Inan et al. and Shin et al. did not directly compare the heart rate from ECG to that of BCG. Instead, Inan et al. (2009) investigated other fundamental factors associated with BCG morphology, such as the relationship of its amplitude to cardiac output while Shin et al. (2009) investigated blood pressure estimation mainly. This is possibly due to the fact that the heart rate of BCG and of ECG have been evaluated in previous studies and were found to have very small errors.

Recent studies have begun to test devices on target populations, such as people with a history of HF. For the weight scale form factor, Giovangrandi et al. (2012) presented preliminary validation results from a study involving four people with a history of HF. Participants' ECG and BCG were recorded using the device over several days at a heart clinic. The results indicated that the BCG obtained from people with heart failure contained more noise compared to that measured from four healthy adults. It was speculated that this was due to movement artifacts and unbalanced posture caused by weaknesses associated with heart failure. Users' perception of the device was reported to be very positive due to the device's simplicity and ease of use.

The authors have expanded the concept of the weight scale to an instrumented floor tile that can be situated anywhere in the home and can unobtrusively monitor weight, heart rate and SBP while the user is performing ADL, such as tooth brushing or hand washing. Stainless steel electrodes were attached on top of the tile to collect F-ECG and load cells were installed on the bottom to collect BCG and body weight. The components of the processing circuit were adopted from previous work, specifically by Inan et al. 2009; Casas et al. 2009. The performance of our prototype in extracting heart rate and estimating SBP was evaluated through trials with 52 healthy adults. Heart rate from BCG was compared to that of ECG and the error was found to be 0.25 per cent (± 0.2 per cent) while SBP estimation had 8.4 per cent error, as previously mentioned. A prototype performance was found to match the works that the design was adapted from, namely Inan et al. (2009) and Shin et al. (2009); however, the change in form factor so that these measures can be captured using a floor tile instead of a weight scale is an important accomplishment of our work. The instrumented floor tile is currently being evaluated through trials with 40 older adults with a history of heart failure by performing a variety of ADL in a home-like environment.

2.5 Summary

The standing-form factors discussed in this section were the weight scale and floor tile. Both forms are able to detect weight, heart rate and in some cases, SBP. BCG is the physiological signal that enables the collection of heart rate and SBP. Methods of obtaining BCG were presented as well as the R-J interval method to combine BCG with ECG to estimate SBP. As they are intended to collect data from people who are standing, standing-form factor devices can be incorporated into virtually any environment and are able to unobtrusively collect important physiological data. This allows placement in areas where people are likely to perform daily activities several times a day, such as in front of the bathroom or kitchen sink. As a zero-effort technology, this form factor can especially be useful for people with chronic conditions, such as heart disease or dementia, since the device will measure vital signs every time the person is on it (i.e. several times a day) and requires no explicit input from the person using it (i.e. he or she does not need to do anything out of the ordinary for his or her vitals to be measured). Solutions to challenges in obtaining these measurements, such as motion artifacts, are currently being investigated.

3. Bed

The bed is a widely used form factor for ambient physiological monitoring. Depending on what sensors are installed, a bed can be used to unobtrusively monitor several physiological traits and corresponding conditions, including sleep health by recording data related to sleep apnea and movement during sleep; cardiovascular disease by measuring weight; and/or detection of wandering with people who have dementia. In terms of signal acquisition, a bed has the advantage of reduced movement artifacts because most likely the occupant will be relatively still when sleeping as compared to other sensors, such as standing on a tile, and resulting in a cleaner signal. A bed is also able to capture data over long periods of time as people often spend several hours in bed whereas periods of recording time for other devices are often in the order of minutes or even seconds.

3.1 Acquisition of BCG

Similar to the floor tile, the majority of research on ambient physiological monitoring using a bed focuses on BCG acquisition. Compared to the floor tile, a bed has more locations where BCG-related sensors may be installed, including on top of the mattress, on the bottom of the mattress, on bed slats and under the bed frame legs. While the mattress provides a good contact surface for collecting data, load cells cannot be used because of the mattress' smooth and flexible surface. Instead of load cells, flexible force transducers, such as strain gauges, are used to detect BCG and respiration signals. A strain gauge is a long, thin wire compacted into a small area. When the surface the gauge is mounted on deforms, the wires inside the gauge also deform and their resistance changes, thereby producing change in the current or the voltage. In bed applications, it is common to use a grid of strain gauges mounted in a pressure pad that is placed on top of the mattress, under the sheet. Using a pressure pad allows for simultaneous collection from multiple points, enabling more accurate representations of a person's sleep pattern.

Other sensors that have been used to detect pressure include air-pressure sensor, hydraulic sensor, air mattress, optical sensor array, polyvynildenefluoride (PVDF) (i.e. electrical foil that converts pressure to electrical energy) and electromechanical film (EMFi) (Bruser et al. 2011; Willemen et al. 2014; Su et al. 2012; Mack et al. 2009; Arcelus et al. 2007; Shin et al. 2008; Friedrich et al. 2010; Watanabe and Watanabe 2004). The type of sensor installed on the bed and the location of the sensor on the bed determine which parameters are monitored. The physiological parameters associated with each sensor are shown in Table 1. Analog circuits are used

Table 1. Types of sensor used on the bed and the associated physiological signals monitored.

Type of Sensor	Installation Location on the Bed	Physiological Signal Monitored
Strain gauge	Middle of a bed slat	Respiration signal, BCG (Bruser et al. 2011)
Load cells	Under bed legs	Weight, BCG, movement, occupancy (authors); BCG, occupancy (Shin et al. 2008)
PVDF, EMFi	On top of the mattress	Respiration signal, BCG (Shin et al. 2008; Friedrich et al. 2010)
Air-pressure, hydraulic sensors	On top of the mattress (Willemen et al. 2014; Su et al. 2012), underneath the mattress (Watanabe and Watanabe 2004; Watanabe et al. 2005)	Respiration signal, BCG (Su et al. 2012; Willemen et al. 2014) (In the paper by Watanabe et al. BCG is indicated as P-CG, cardiogram obtained by pressure sensor)
Pressure pad	Underneath the mattress	Respiration signal, movement, occupancy (Arcelus et al. 2007)
Optical sensor array	On top of the mattress	Respiration signal, BCG (Bruser et al. 2012)
Accelerometer array	Blanket on the bed	Respiration signal (authors)

to amplify these signals and consist of similar configurations as the ones used for weight scales and floor tiles that were discussed in 2.1.

The use of pressure sensors to detect heartbeat is also being investigated. One example is the work done by Shin et al. (2008) that compared the effectiveness of three sensors (air-pressure sensor, load cell and EMFi) in heartbeat detection. The air-pressure sensor underneath the mattress and load cells under the bed's legs were found to be similar in performance and outperformed EMFi in detecting the heartbeat. While the performance of the other sensors has not yet been compared to one another, all the sensors have been reported as being successful in detecting BCG from the trials conducted by the developers.

One of the main objectives of signal processing for the bed-form factor is to detect the peak of BCG more effectively, before separating a respiration signal from BCG. Some of the algorithms that have been used are:

- Use of BCG template (i.e. ensemble average of BCG) to acquire the heart rate from BCG (Shin et al. 2008)
- Heartbeat detection using K-mean clustering (Rosales et al. 2012; Bruser et al. 2010)
- Use of a second order band-pass filter followed by detection of local peaks to separate BCG and respiration signal (Willemen et al. 2014; Mack et al. 2009)
- Use of locally projective noise reduction (LPNR) algorithm, which is non-linear, to separate BCG and respiration signal (Yao et al. 2014).
- Artificial intelligence using a Bayesian approach to obtain heart rate without training data (2013)

The above algorithms branch into two categories—the heartbeat detection algorithms and decomposition algorithms (i.e. algorithms to separate respiration signal and heart signal). Within the category of heartbeat detection algorithms, the template-matching algorithm developed by Shin et al. (2008) uses templates to detect peaks of heartbeats from BCG. Shin et al. (2008) tried this method on three different sensors (i.e. air pressure sensor, load cells underneath the bed legs and EMFi). The method produced high accuracy for load cell and air pressure sensor but generated more false detections for EMFi. The group reported that the algorithm needs to have a template for each sensor and works well when the signal is stable and the each wave is clear in each heartbeat. Rosales et al. (2012) performed k-mean cluster method to detect heart rate from BCG. The advantage of this algorithm is that unlike the method developed by Shin's group, signals that do not exhibit clear self-repeating patterns can be used to detect the heartbeat peaks. The algorithm was tested on four subjects—analysis on three subjects showed accuracy between 97.7 per cent and 100 per cent while the remaining subject had movement artifacts, which reduced the accuracy to about 83 per cent. In a separate work, Bruser et al. (2010) developed a BCG heartbeat-detection algorithm based on k-mean cluster algorithm than other novel band-pass filtering techniques to detect heartbeat peaks. Results from 17 subjects showed a mean error less than one per cent during the period where there was no severe movement artifacts. The group improved the algorithm by implementing a Bayesian approach (i.e. the algorithm self-adjusts as the signal is being collected) to remove the training stage that non-Bayesian k-mean clustering (and many other methods) require for each user in order to implement the algorithm. The resulting k-mean Bayesian algorithm was tested on 33 subjects, including 25 insomniacs, and resulted in 72.69 per cent of the signal being usable, with a mean error of 0.78 per cent (2013). When the results were compared between the healthy adults and insomniac subjects, the healthy adults had more coverage with less error; healthy adults had the coverage of 84.53 per cent with a mean error of 0.61 per cent whereas the insomniac subjects had the coverage of 68.9 per cent with a mean error of 0.83 per cent.

The second category of algorithms is those that separate the respiration signal and heart signal, which are both embedded in BCG signals. Willemen et al. (2014) and Mack et al. (2009) separated the respiration signal and heartbeat signal based on BCG using a second-order digital Butterworth band-pass filter with a bandwidth of 0.05 to 0.5 Hz for respiration signal and the bandwidth of 2 to 15 Hz for the heartbeat signal. The decomposition was considered to be successful and the resulting data were used for further analysis; however, the method is ineffective if the frequency component of the signal goes out of the range of the band-pass filters (e.g. if the person hyperventilates and respiration goes above 0.5 Hz). As well, if the frequency components of the signals overlap, it is not possible for a simple linear band-pass filter such as this one to separate them. Yao et al. (2014) used a locally projective noise reduction (LPNR) algorithm, which is non-linear, to overcome this problem. A LPNR algorithm filters the signals more selectively by projecting the data on to the nullspace to estimate the noise and remove from the noisy signal. In this case, noise and the filtered signal are both useful since they are respiration signal and BCG. A provisional qualitative analysis was done by Yao et al. (2014) and the results indicated that the LPNR algorithm performed better than a conventional band-pass filter.

The works presented above demonstrate how algorithms are evolving as the later studies built on the results of previous works. It is anticipated that this evolution will continue and these types of algorithms will continue to increase the ability to extract accurate heart beat and respiration data from a bed-form factor.

3.2 Current Research

While the use of a bed as an ambient physiological monitoring system is fairly new, there have been some tests on healthy adults and people who have sleep disorders, such as those listed in Table 2.

Table 2. Examples of trials comparing the heart rate from BCG obtained from an ambient monitoring bed to conventional ECG obtained from chest.

Study	Number of Subjects	Hours Recorded	Coverage[1]	Error Compared to ECG	Method Used
(Bruser et al. 2013)	33 (8 normal, 25 insomniacs)	13167 (6:39 hours per person)	72.69%	0.78%	Strain gauge on a bed slat
(2014)	46 (10 normal, 36 sleep clinic patients)	Overnight recording for each patient	54.07%	1.27%	EMFi underneath the bed poles or underneath the bed sheet
(Jung et al. 2014)	20 (10 normal, 10 obstructive sleep apnea (OSA))	145 hours (7.3 hours per person)	Sleep efficiency[2]: 91% Sensitivity[3]: 85.2%	Accuracy: 97%	PVDF film on top of the mattress
(Aubert and Brauers 2008)	58 (11 normal, 19 OSA, 6 insomnia, 22 other sleep disorders)	740	83.2%	1.25 beats per minute (BPM)	EMFi underneath the mattress

[1] 'Coverage' is the per cent of useful BCG that is free of movement artifacts. Note that the definition of coverage varies between researchers and between the conditions that are being monitored, leading to exacerbated differences in percentage of coverage. Regardless, the reported errors are acceptably low, showing that BCG from the bed can be used to collect the heart rate.

[2] 'Sleep efficiency' is the ratio of time spent asleep (total sleep time) to the amount of time spent in bed.

[3] 'Sensitivity' is true positive heartbeat (i.e. correctly detected heartbeats) divided by sum of true positive and false negative (i.e. not detecting a heartbeat when there is one) heartbeats.

Within the past few years, a number of studies that use a bed were conducted on people with sleep disorders to evaluate the functionality of their devices. The types of disorders that have been examined include insomnia, obstructive sleep apnea (OSA) and other sleep disorders (Jung et al. 2014; Aubert and Brauers 2008). Bruser et al. (2013) compared 11 healthy adults with 25 people with insomnia and found that the latter had statistically significant lower values of coverage and higher variances of error compared to healthy adults (i.e. 68.9 per cent and 84.53 per cent of coverage for people with insomnia and healthy adults, respectively; 1.23 per cent and 1.61 per cent of 95th error percentile for people with insomnia and healthy adults, respectively). Jung et al. (2014) compared 10 healthy adults with 10 people with OSA and found that there were no statistically significant differences between people with OSA and healthy adults in accuracy and coverage (indicated via sleep efficiency and sensitivity in the original paper); the people with OSA had lower accuracy and coverage by only 1 to 2 per cent. In their study, Aubert et al. (2008) compared 11 healthy adults with 19 people with OSA, 6 people with insomnia and 22 people with other sleep disorders. This study reported 88 per cent, 74 per cent, 88 per cent, and 79 per cent heart rate and 91.5 per cent, 80 per cent, 89 per cent, and 81 per cent respiration rate coverages for healthy adults, people with OSA, people with insomnia and people with other sleep disorders, respectively. In all these studies the coverage of the people with sleep disorders was lower than that of healthy adults. Bruser et al. (2013) and Aubert et al. (2008) noted coverage (i.e. periods where the physiological signal is acquired properly; described in Table 2) decreased by 10 to 15 per cent for people with a sleep disorder. On the contrary, according to Jung et al. (2014), coverage is not affected by the presence of any sleep disorder. The contradicting results indicate the need for further investigation in this area. In terms of heart rate, these three studies are in agreement that detection from people who have a sleep disorder was slightly lower but almost equal to that of healthy adults within periods of coverage (Jung et al. 2014; Aubert and Brauers 2008). None of these studies attempted to identify if the subject had a sleep disorder based on the data, as algorithm development was not yet sophisticated enough to make such an assessment. However, investigation on the state of sleep (discussed below), can be used to detect sleep disorders in the future.

Research on detecting adverse events using ambient physiological monitoring has begun to appear. Bruser et al. (2013) developed and tested seven machine-learning algorithms designed to distinguish atrial fibrillation based on BCG recordings from the bed on 10 subjects with atrial fibrillation. The best algorithm had Matthews correlation coefficient, meaning sensitivity and mean specificity of 0.921, 0.938, and 0.982, respectively. Based on these results, the group concluded that their algorithm could be used in a home environment to detect atrial fibrillation events. However, this has yet to be tested in real-world applications. Nevertheless, this study is the first of its kind to interpret abnormal BCG recording to detect adverse events. Other than heart rate, the possibility of collecting respiration rate from the BCG was evaluated by Willemen et al. (2014), Mack et al. (2009) and Yao et al. (2014) as discussed in 3.1. This research showed that respiration signal could be obtained by using pressure pads or air pressure sensors installed on the bed.

There are several small-scale studies involving people with abnormal conditions, such as cardiac disease or older adults within the test subjects, such as (Rosales et al. 2012; Su et al. 2012; Friedrich et al. 2010). However, the sample sizes for these studies were less than 10 and there was no comparison between the cohorts involved; thus, these works are not discussed here.

The clinical significance of detecting weight changes in real-world scenarios using a bed has not yet been evaluated; however, small-scale controlled trials in laboratory conditions involving people with a history of heart failure are being conducted by the authors. Preliminary data analysis indicates that the bed is successful at capturing the weight of the person lying on it. The measurement of the weight could assist the people with heart failure in short-term and long-term tracking of their weight, which could result in more comprehensive monitoring of their condition as well as in detecting fluid retention (Refer to Section 1.2 for more details).

Currently, sleep quality is usually assessed using polysomnography (PSG), which is a comprehensive method for evaluating the state of sleep by measuring ECG, electroencephalogram (i.e. electrical activity of the brain) (EEG), respiration and other signals. PSG requires the person being assessed to wear multiple instruments while sleeping, which can be burdensome, can affect sleep quality and is not practical in a home environment. Simplifying this process of monitoring sleep motivated researchers to investigate by using BCG to assess sleep. Several groups conducting research on the feasibility of classifying the sleep state (e.g. awake, REM, non-REM states) using BCG have all attested that it can be used (Kortelainen et al. 2010; Jung et al. 2014; Paalasmaa et al. 2012; Mack et al. 2009; Migliorini et al. 2010). Mack et al. (2009) claimed that BCG estimation of sleep state outperformed the actigraphy, which is a conventional method for assessing the sleep state as part of PSG.

Bed occupancy is an important metric that could be used to detect possible wandering of people with dementia. The topic was not discussed in the studies so far, possibly due to the fact that people with sleep disorder were focused. However, bed occupancy could conceivably be detected using signals such as the ones described above. For example, movement artifacts or no signal due to the absence of the user from the bed could be valuable in determining if someone is in the bed, or is in the process of getting into or out of bed.

While unobtrusive blood pressure measurement have been investigated for standing-form factor applications, these have not yet been developed for a bed. The majority of studies focus on detecting sleep disorders and therefore focus on respiration and movement rather than on other physiological measures, such as blood pressure. However, recent studies reveal that changes in nighttime blood pressure is a significant indication of increased risk of cardiovascular morbidity and mortality (Calhoun and Harding 2010). In light of this, the measurement of blood pressure using a bed has great potential to benefit people with cardiovascular disease and should be investigated in all future developments.

A small number of studies used capacitive-coupled electrodes (CC-electrodes) (refer to 4.2) to detect ECG from a bed through fabric. Wu et al. (2008) placed large conductive fabrics beneath a sheet on the bed to monitor ECG while the occupant was a sleep. In Wu et al.'s experiment, the bed was tested with seven male and one female lying in different positions (e.g. supine, right-turned, left-turned and prone positions), and an average heart rate detection accuracy of 95 per cent was achieved. Kato et al. (2006) developed a similar system to detect infants' ECG on a bed and they designed the locations and shape of the electrodes based on the pressure that a baby exerted on the bed. In Kato et al.'s experiment, 4 infants, aged 10 to 133 days, were recruited to test the bed. It was reported that a clear ECG waveform was collected from a 115-day-old baby, while the system failed to capture the 10-day-old baby's ECG.

3.3 Summary

The bed-form factor has advanced further in terms of algorithmic development and clinical evaluations compared to those of the standing-form factor and the sitting form factor, which are discussed in Sections 2 and 4 respectively. This is because of greater availability of stable data and well-defined future clinical applications. In this section, different types of sensors and their effectiveness are examined in addition to the algorithms used for detection of heart rate and respiration rate. Thus far, most bed prototypes have been tested on healthy adults with only a few studies evaluating the devices with people who have a sleep disorder. The benefits of bed applications could be expanded to other populations, such as people with heart failure and dementia by monitoring conditions such as weight and wandering. Being a zero-effort technology, the bed-form factor greatly increases the usability for all three chronic diseases. This is especially true for sleep health assessment since the conventional way of assessing sleep health (e.g. polysomnography) requires multiple obtrusive equipment to be worn by the person being monitored while a sleep. The bed-form factor will require no equipment at all. Overall, the bed is a promising form factor to collect BCG in order to assess sleep health, heart failure, dementia and other morbidities.

4. Sitting-Form Factors

Another form factor that is used for ambient physiological monitoring is a sitting-form factor, such as a chair or a toilet seat. Similar to the bed-form factor, a chair provides a prolonged period of time with a relatively stable posture, which minimizes movement artifacts during monitoring. Additionally, toilet seats can provide good skin contact. To date, most chair devices have been used to collect BCG; PPG is collected in some chair prototypes and ECG is measured through clothing in some prototypes using capacitive-coupled electrodes (CC-electrodes). Either BCG or PPG can be used to measure the heart rate while a combination of ECG and PPG can be used to estimate blood pressure by utilizing a parameter called pulse arrival time (PAT). The same principles that show a correlation between heart rate of BCG and that of ECG for the bed-form factors discussed in 3.1 are applicable to sitting-form factors. The reader can refer to Section 1.2 for information about ECG, BCG and PPG.

The rest of this section focuses on estimating blood pressure by using pulse arrival time and capacitive coupled electrodes work as well as discussing sitting-form factor applications.

4.1 Blood Pressure Estimation Based on Pulse Arrival Time (PAT)

Pulse arrival time (PAT) is a physiological parameter that is indirectly related to blood pressure (discussed in more detail below) and has been used to estimate blood pressure. PAT is the time interval between the R-wave of an ECG and arrival of the arterial pulse to periphery, which is indicated by PPG (ECG and PPG are discussed in Sections 1.2.1 and 1.2.3, respectively) (Rassaf et al. 2010). Typical PPG sensors require contact with the skin and is usually measured from the fingertips. The PAT method has been investigated since early 1970s; so it has been studied for much longer and in greater depth compared to the R-J interval method (discussed in Section 2.2), which first emerged in 2008 (Inan et al. 2008).

It is known that blood pressure is proportional to pulse transit time (PTT), which is measured by recording the time taken by an arterial pulse to travel between two points of the same artery separated by a known distance (Zhang et al. 2011a). The relationship between blood pressure and PTT is well established by the Moens-Korteweg equation where PTT can be used to measure blood pressure (Muehlsteff et al. 2013). A simpler but less accurate alternative to PTT is to measure PAT and use it to estimate blood pressure. PAT comprises two quantities: pre-ejection period (PEP) and PTT. PEP is the time from when initiation of left-ventricle contraction begins to the point blood goes out of the ventricle and into the arteries. PEP measures can change because of multiple factors, including changes in blood pressure, body position, etc. Thus measuring PAT gives an indication of PTT; however, this measurement can be influenced by changes in PEP (Zhang et al. 2011b). Despite these limitations, PAT has received much attention in estimating the blood pressure because of its non-invasiveness and simplicity. PTT requires the arterial pulse to be detected (e.g. using PPG and arm cuff) from two known locations that are at different distances from the heart. Data from these locations can be used to calculate how fast the pulse travels. PAT requires the measurement of ECG and only one arterial pulse (Geddes et al. 1981). Both methods require two contact points. However, ECG and PPG can be combined in one device and attached to one location of the body, whereas PPT must have two sensors attached to the body and the distance between the sensor must be known when worn. Finally, having ECG and PPG will provide more information regarding one's health than two identical signals (i.e. arterial pulse wave).

Work has been done to improve this drawback by estimating PEP via addition of BCG or impedance cardiogram (ICG) (i.e. measurement of change in resistance across the thorax region of the body due to blood flow and volume change 2009) and incorporating it into the blood pressure estimation (Wong et al. 2011; 2012).

While it has been shown that using PAT enables the estimation of SBP, it still appears to be ineffective in estimating diastolic blood pressure (DBP) (Kim et al. 2008). One of the recent advances in the PAT method has enabled the estimation of DBP by using PAT and additional parameters. By incorporating heart rate and arterial stiffness index into a multiple regression analysis, both SBP and DBP estimations from PAT were improved (Baek et al. 2010). Lastly, movement artifacts cause a degradation of estimation quality in the PAT method. Sola et al. (2009) developed an algorithm to reduce this effect by extracting information from the whole anacrotic phase (i.e. the first concave shape upswing phase of PPG (Button 2005)) using hyperbolic tangent parametric estimator (TANH) instead of using a particular characteristic point of the waveform. The function of the TANH is to estimate the shape of the anacrotic phase and match the noisy PPG signal to find the location of anacrotic phase. Compared to the commercial device, the method required five times less heartbeats to make a reliable PAT measurement (Sola et al. 2009).

The PAT method has been evaluated by a few studies. For instance, the study by Rassaf et al. (2010) evaluated the performance of PAT method post-exercise and in different postures. PAT method successfully estimated the rising systolic blood pressure at the beginning of the exercise but did not reflect pressure properly during the recovery period. The PAT method was not successful in estimating SBP during posture changes, possibly due to changes in PEP (Rassaf et al. 2010). Muehlsteff et al. (2013) evaluated the detection of impending fainting, also known as syncope caused by a sudden drop in blood pressure and claimed that the PAT method can reliably be used to detect this condition. In general, PAT is a good indication of change in PEP and PTT. When PEP is not affected (e.g. when the subject is stationary), changes in PAT appear to reflect changes in PTT and thus, blood pressure. However, in cases where PEP changes simultaneously with PTT (e.g. change of position or exercise (Wong et al. 2011)), the estimation of blood pressure using PAT gets complicated.

In terms of technical improvements, Baek et al. (2009) recently enhanced conventional PPG so that the signal can be measured through clothing without the need of a direct contact with the skin (Baek et al. 2009). In validation trials with five healthy subjects, the correlation coefficient between the method and the gold standard (i.e. PPG through skin contact) was higher than 0.9 for all subjects. Baek and colleagues also used this method in a trial involving five different subjects to calculate PAT and validated that PPG measured through clothing could be used as a surrogate for the PPG measured from direct skin contact (Baek et al. 2012). This study is discussed further in Section 4.3.

4.2 Capacitive Coupled Electrodes (CC-Electrodes)

As described in 1.2.1, ECG is usually collected by attaching electrodes directly on to the skin. Electrolytic gel at the attachment point creates a low impedance pathway for ECG propagation into the sensing instrument, resulting in excellent signal quality. However, in the context of monitoring in home settings, this conventional wet electrode system has several limitations. First of all, the wires and gelled electrodes are obtrusive to the user's daily activities; second, the application of wet electrodes to the skin requires training and careful placement; thirdly, long contact times with adhesive and gel may irritate the skin.

In the light of the limitations of wet electrodes, the use of capacitive coupled electrodes (CC-electrodes) is proposed. Unlike gelled electrodes, which rely on galvanic contact, CC-electrodes are insulated from skin by dielectric material (i.e. air gap, cloth, etc.). The electrode, dielectric material and skin form a capacitor, so that an ECG signal is able to propagate from the skin to the electrode through a capacitive coupling (Fig. 2). However, if capacitive interface is a high impedance component, this will degrade the signal strength and requires high input impedance instrumentation system to extract the signal. Thanks to the advancements in semiconductor technology, the availability of high impedance and low-noise analog front-ends (i.e. processing components) enables the implementation of capacitive ECG systems. Readers may refer to Chi et al. (2010) to learn more about how to extract an ECG signal from CC-electrodes.

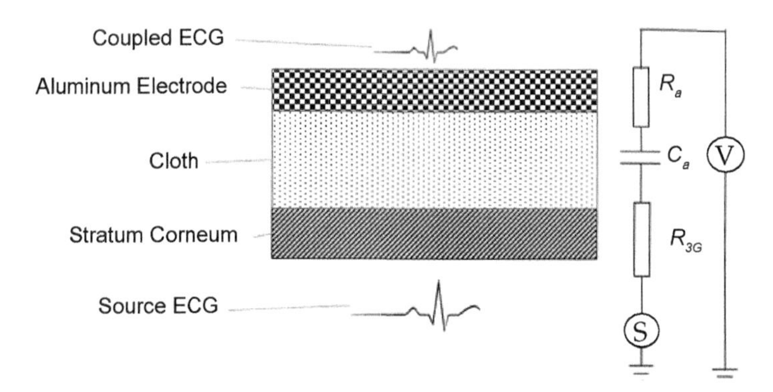

Fig. 2. Simplified circuit model for ECG acquisition through CC-electrodes. R_{sc} represents the impedance between the heart (i.e. signal source) and skin surface; C_a is the equivalent capacitance of electrode-skin interface (in the orders of *pF*); R_a is the input resistance of the measurement instrument.

Some research focuses on deploying CC-electrodes into daily living environments to achieve unobtrusive and long-term monitoring of cardiac health of occupants. Some techniques were adopted to enhance signal quality when CC-electrodes were embedded into the chair. Lim et al. (2009) used Driven-Right-Leg (DRL) (i.e. a technique commonly used to reduce the 60 Hz power line noise in the signal) to reduce noise—with a gain of 100, DRL reduces the power line noise by 30dB. Additionally, Kim et al. (2006) showed that doubling the coupling capacitance of DRL (from $1nF$ to $2nF$) brought 5dB signal to noise (SNR) ratio improvement.

Although the CC-electrode is a versatile way of measuring ECG through fabric, several challenges exist in using the module. As a capacitive ECG system has high signal sensitivity, it becomes susceptible to noise, notably 60 Hz noise from the power lines running to wall plugs and motion artifacts. Additionally, in a non-constrictive setting, new system variables are introduced; for example, user posture, clothing material and electrode-embedding scheme will affect the electrode-skin interface, which increases the dynamic range of the signal output.

4.3 Current Research

Though several devices have implemented one or more of BCG, PPG or CC-electrodes, to date, they have only been evaluated with pilot trials involving healthy adults. A toilet seat that has electrodes and PPG sensors on it was used to estimated SBP on five healthy subjects; the correlation coefficient between the estimated and the true SBP was 0.766, which represents the ability to provide approximate estimation of SBP with some uncertainty (Kim et al. 2006; Baek et al. 2010). In other words, while the methods do not give a precise SBP value, but when there is a significant change in SBP, the method reliably detects the change. Baek et al. incorporated a system that detects PPG through the fabric in an office chair while using the electromagnetic film (EMFi) to obtain BCG and CC-electrodes to obtain the ECG (Baek et al. 2012). The group evaluated the effectiveness of the device to estimate the blood pressure in five healthy adults and the correlation coefficient between the estimated and true SBP was 0.755. Researchers claim that the chair collected ECG properly from the CC-electrodes as well as the BCG from the EMFi (Baek et al. 2012). Arcelus et al. (2013) implemented a CC-electrode design that can sense ECG on a dining chair, with heart rate inference accuracy being around 98.8 per cent as compared to the standard ECG measured from the chest when tested with a female subject sitting on the chair. This design is currently being evaluated through pilot trials with 40 participants with HF. Other examples include chair devices that measure only BCG. Karki and Lekkala (2009) and Akhbardeh et al. (2007) were successful in obtaining BCG and extracting the heart rate.

Note that the toilet device by Kim et al. (2006) and the office chair by Baek et al. (2012) measure PPG on the thigh of a person, which is atypical considering that PPG is usually collected from the fingertip. The developers did not explicitly comment on whether PPG from the thigh can be a surrogate for the PPG measured from the fingertip. While the developers appeared to be successful in finding a similar pattern of PAT, it is yet to be verified whether PPG from the thigh can be a substitute for the PPG from the fingertip.

Thus far there have has been two studies that involve subjects with a history of cardiovascular disease, but the sample size was too small (i.e. three and six subjects) for a comparative or significant validation of the devices to be made (Junnila et al. 2006).

4.4 Summary

This section examined sitting-form factors used to achieve the ambient physiological monitoring, including a chair and a toilet seat. Apart from measuring BCG, devices that use a sitting-form factor are able to use PPG and ECG to measure the heart rate and estimate the blood pressure. The concept of blood pressure estimation, based on the parameter resulted from the combination of ECG and PPG, namely PAT, as well as acquiring ECG through clothing by using CC-electrodes, were discussed. Measurements such as these could benefit people with chronic health conditions as regular heart rate and blood pressure measurements can help monitor health and identify the possible onset of adverse events. By performing these measurements using zero-effort technology, the sitting-form factor monitoring also avoids the confusion and inefficiency caused by incorrect operation or forgetting the measurement schedule. Sitting-form factor devices exhibit promising results and are likely to be part of the suite of ambient physiological monitoring devices in the near future.

5. Activity Monitoring

An important aspect of the ambient physiological monitoring paradigm is activity monitoring. Activity monitoring for health care applications can support independent living and aging in a place for older adults by detecting activities of daily living (ADL), monitoring health and well-being status based on the detection, and indicating if support is necessary (Chen et al. 2012; 2014).

An ability to perform ADL is required to live independently and any deterioration in performing ADL may indicate that adverse condition (e.g. physical and/or cognitive illness) is taking place (Katz et al. 1963; Reisberg et al. 2001). In relation to dementia, the stage of dementia can be assessed by examining how ADLs are performed. For example, one way occupational therapists assess the severity of dementia and level of cognitive support required is to examine how well a person with dementia performs a specific task in a kitchen (2007). Gait (and in particular, walking speed) seems to be another indicator of the onset of dementia. Waite et al. (2005) reported that based on a six-year study involving 630 older adults, aged 75 or over with slowing gait and motor functions were most likely to be diagnosed with dementia at the end of the study and had the highest mortality rate. Based on facts such as these, activity monitoring aims to identify changes in activity that may suggest a decline in health, so that appropriate interventions are taken to help the person being monitored to maintain his or her wellbeing and independence to the greatest possible extent.

Activity monitoring is achieved through collecting and analyzing data from sensors placed in a living environment. Sensors can range from vision-based (i.e. using a video camera and resulting video data) to ambient-type sensors (e.g. RFID tags and simple switches) and sensors can be wearable or embedded into an environment (Chen et al. 2012). Sensors installed for active monitoring often produce discrete outputs; many of them are binary outputs (i.e. on or off), for example, magnetic sensors attached to the door of a house will indicate if the door is open or closed. Sensors are sampled at regular intervals with timestamps for each sample, allowing the person looking at the data to know how long an event took place.

Multiple sensors are installed around the home to record an array of data, namely data obtained from one or more sensors over a period of time. The data can then be processed to identify what and how an activity is being performed. This process is called activity recognition. Building the necessary model to achieve recognition (called 'activity modeling') is a complex process and often employs techniques from the field of machine learning, where a set of methods is applied to detect patterns in data, use the patterns to predict future data, or perform other kinds of decision making (Murphy 2012).

Krishnan and Cook (2014) conducted a six-month long trial in three homes with three healthy participants. Their algorithm was trained (i.e. 'taught' what normal-activity behaviours looked like) in the same apartments with the same participants before the trial began. The researchers evaluated six different algorithms and tuned the algorithms' parameters to achieve optimal results for each. The best performing algorithm achieved 68 per cent accuracy compared to human annotated data. Interestingly, more than 50 per cent of the sensor data was categorized as 'other', which means that more than 50 per cent of the sensor data pertained to related activities that were not included in the analysis. The researchers pointed out that solving issues related to this 'other' category was an important future task.

This study is one of the first real-world applications and gives an idea of the current state of activity-recognition research, namely, laboratory-based techniques are slowly being transitioned into real-world environments. Kaye et al. (2011) and Akl et al. (2015) are examples of studies that focus on activity monitoring but do not involve activity recognition (i.e. using a machine learning technique to recognize a specific activity, such as handwashing based on the sensor data). Kaye et al. (2011) conducted a 33-month long study involving 233 healthy adults who were 80 years old or more. Several sensors were installed in the participants' residences, including wireless infrared sensors to detect the motion, modified infrared sensors with a narrow field of view (i.e. sensors fired only when the subject is directly below the sensor) and wireless magnetic contact sensors to detect opening and closing of the doors. The parameters that were measured included walking speed, nighttime activity, the number of walks per day and time out of the home. The participant also filled out a weekly questionnaire on their condition and any adverse events, such as a fall, that may have occurred during the week. There have been several interesting findings from this work. For example, it was found that the mean walking speed was found to be lower when it was measured by ambient sensors than during an assessment by a human. This result supports the notion that people (consciously or unconsciously) modify their behaviour during assessments. The work by Kaye et al. constitutes the foundational work regarding data collection from older adults living in real-world environments.

Akl et al. (2015) used data collected by Kaye et al. (2011) to investigate whether cognitive decline could be detected from ambient sensors. Based on data from 97 of the subjects, Akl et al. (2015) applied machine-learning algorithms and other signal processing techniques to classify the presence of mild cognitive impairment (MCI) in the subjects. This was accomplished using a number of parameters regarding variations in walking speed, weekly number of outings, amount of daily activity, age and gender. Based on the assessment by Kaye et al. at the beginning, through the duration and at end of the study, two of the 10 male subjects and 16 of the 87 female subjects had MCI by the end of the study. The most important features for detecting the presence of MCI in older adults were trajectories of weekly walking speed, coefficient of variation of the walking speed, coefficient of variation of the morning and evening walking speed and the subjects' age and gender. Based on a support vector machine (SVM) algorithm, detection of MCI in older adults was made with an area under the receiver operating characteristic curve of 0.97, which can approximately translate into a 97 per cent probability of having a correct classification, using a sliding time window of 24 weeks.

The significance of the studies conducted by Kaye et al. (2011) and Akl et al. (2015) is that it is possible to use computers to identify MCI from sensors collecting data in real-world environments. In other words, sensors could be installed in people's homes and conditions like the development of MCI could be autonomously detected if and when they arose. It is expected that research to investigate this potential will be evaluated in the future.

The research on activity monitoring is moving towards its goal, which is achieving early detection of the onset of dementia and other adverse health events. With respect to detecting abnormal activity via activity recognition, Chen et al. (2012) pointed out two main challenges. First, abnormal activity has not been well defined in the field of activity recognition as the applications are diverse and the data and contexts are still relatively scarce. Second, a much larger amount of data is available for normal (healthy) activities compared to that of abnormal activities, making it difficult to train algorithms to recognise adverse conditions. There has also been significant progress in other applications, such as detecting the onset of MCI.

Much research is being done with respect to activity monitoring, in terms of types of sensors, data analysis and monitored ADL. The reader is referred to publications, such as Dodge et al. (2012); Chen et al. (2012); Chen et al. (2014); Hong and Nugent (2011) as examples of this work.

5.1 Summary

Research in the field of activity monitoring is progressing quickly and may soon enable early detection of conditions like the onset of dementia. So far, the bulk of research has been done in laboratory environments to recognize scripted and pre-segmented activities based on machine-learning techniques. Current research on activity monitoring is slowly shifting from the constrained laboratory environment to unconstrained, real-world settings. Similar to the physiological monitoring techniques discussed in previous sections, activity monitoring has an enormous potential to help people identify adverse events promptly so that appropriate interventions are put in place to maintain their quality of living to the highest possible degree. Moreover, the process is achieved with zero-effort technologies, which provides benefits not only in the clinical aspect but also in the usability and compliance aspects. In other words, while their activity is being monitored regularly, the person being monitored does not have to actively participate in or even think about performing the monitoring.

6. Discussion and Future Work

Physiological monitoring at home has enormous potential regarding the tracking and management of health conditions, which may translate into significantly better health outcomes, particularly people with chronic conditions. Unobtrusive physiological monitoring devices and related principles have been discussed throughout this chapter. The field of ambient physiological monitoring is still in its infancy and needs to be further developed and refined, both technically and clinically, before it is ready for implementation in day-to-day clinical practice (Peetoom et al. 2014). A number of improvements are expected to take place in the next few years, including technical developments to make the current devices robust in real life situations, involvement of people with physiological conditions and longitudinal clinical trials at the homes of the target populations.

6.1 Technical Development

There are a number of aspects that must be tackled to facilitate the use of ambient physiological monitoring devices in real-world environments. These include reduction of movement artifacts, disambiguation of multiple people present at home, detection of adverse events, increasing unobtrusiveness via incorporation of state-of-the-art sensors and networking of the ambient monitoring devices.

6.1.1 Movement artifacts

Most vital sign-monitoring devices have problems with movement artifacts. While the bed-form factor compensated for this difficulty by having a much longer capture time, thus reducing the per cent of data that includes motion artifacts, resilience to movement artifacts can still improve the quality of data that is captured. BCG, PPG and ECG through CC-electrodes are all very susceptible to movement artifacts at the current stage of ambient monitoring. There are different levels of severity in movement artifacts by way of relatively small movement artifacts being removed while larger artifacts overpower the signal and related data may simply not be usable. Movement artifact removal is commonly achieved by two methods. First, signal processing algorithms can be applied, such as filters and machine learning techniques, to reduce the movement artifact noise. The second method is to incorporate additional sensors or improve the signal and compensate for movement artifacts. Applying the latest sensor technologies and noise reduction techniques will also accommodate certain levels of movement artifacts.

6.1.2 Disambiguation of multiple people

One of the technical challenges that ambient physiological monitoring devices must overcome is disambiguation of multiple people present in the home environment; namely, the system must know who is being monitored at any given time. Most ambient physiological monitoring devices are nonspecific by way of who generated each measurement and the current devices lack the ability to identify individuals when there is more than one person living in a house. For instance, it is common for two people to sleep in the same bed; different individuals may sit on the same chair at different times of the day and more than one person might stand on a floor tile. To date, there have been no studies to examine multiple people using the same ambient vital sign-monitoring devices, namely the bed, chair and weight scale-form factors. Some ideas of distinguishing one user from another can be found in the area of biometric security, in which biometric signals are used as a means of personal identification. Applications from the field of biometric security can be adopted to ambient physiological monitoring to identify who is using the monitoring devices and then process the data according to the user's profile. One example is the personal identification via ECG, which has been a popular modality in the past decade with a significant amount of work done on the topic (Odinaka et al. 2012). Use of the frequency domain features of PPG as the identification key has been also proposed (2008). There have been a few investigations into the disambiguation of users for activity monitoring. For example, the Microsoft Kinect sensor has been used to assess the gait parameters and the height of people being monitored to identify each individual (Stone and Skubic 2011;

Rantz et al. 2013). Austin et al. (2011) also identified different individuals by looking at each person's gait velocity. Results from a study of 20 older adult couples (mean age of 81.8 years) showed that when the gait velocity of the couple differed by more than 15 cm/s, it was possible to identify each subject with the correlation coefficient of 0.877. When the restriction of 15 cm/s was removed, the correlation coefficient was 0.54. The disambiguation of users is an important problem that has not yet received much attention. In the next few years, it is expected that this issue will be examined closely as the ambient physiological monitoring devices are tested in real home environments.

6.1.3 Algorithm to detect adverse events

Algorithms that allow detection of adverse events would be a valuable component of ambient physiological monitoring devices to be implemented in homes of people with chronic illnesses.

Some algorithms related to sleep health monitoring have been developed for bed-form factors that use BCG, namely, several studies developed algorithms to classifying sleep state (Kortelainen et al. 2010; Jung et al. 2014; Paalasmaa et al. 2012; Mack et al. 2009; Migliorini et al. 2010). It is anticipated that these algorithms will advance further to enable preliminary diagnosis of sleep disorders in the near future. Algorithms related to cardiovascular disease have yet to be developed, with the exception of the atrial fibrillation detection by Bruser et al. (refer to Section 3.2). At present, the algorithms used to extract physiological parameters from healthy adults are in the process of maturing; it is expected that over the next few years, methods to detect adverse physiological events will emerge along with their evaluation of devices by their intended users, including physicians and other clinical professionals, as well as the people who are being monitored.

Inan et al. (2014) pointed out that in order for research on ambient physiological monitoring to thrive and accelerate its usefulness, the tools and resources used to perform the research must be made available to the research community, such as the development of open source databases of ambient physiological signals. Indeed, research on ECG have benefitted immensely from the availability of open source physiological signal databases, such as the one provided by the Massachusetts Institute of Technology-Beth Israel Hospital (MIT-BIT) database (http://ecg.mit.edu/). Similar such resources will help accelerate many types of algorithm development.

As the adverse detection algorithms emerge, it is probable that the ambient physiological monitoring devices will be able to detect the onset or exacerbation of diseases, making the technology more effective and useful as a diagnostic, rather than a simple monitoring, tool.

6.1.4 Incorporation of state-of-the-art sensors

One of the areas that ambient physiological monitoring is expected to improve is its usability and unobtrusiveness through the incorporation of progressively advanced sensors. Hard electronic components that are currently used, such as the photodiode and LED light used for PPG, will be replaced with flexible sensors to increase the blending of the sensors into the furniture. Although the flexible sensors, such as electromechanical film (EMFi) does not interfere with a user's activites, its unobtrusiveness can be improved by using newly developed sensors to make the sensor-furniture assembly completely seamless and blended. Examples of the newly emerging sensors include flexible photo-transistor and organic light emitting diode (OLED) for PPG sensing (Chang et al. 2013; Lee et al. 2014) and carbon nanotube (CNT)-based flexible film strain sensors for BCG and respiration sensing. CNTs can measure strain 50 times greater than the conventional strain gauge being incorporated into a fabric (Yamada et al. 2011). However, these sensors are still in development stage and need to be proven as durable and reliable prior to being incorporated in a consumer-ready product (Zheng et al. 2014).

6.1.5 Ambient physiological monitoring device networks

Most of the seminal works discussed in the chapter remain on only one form factor. Forming a network of ambient physiological monitoring devices is a future task that would enable ambient physiological monitoring-form factors to share and aggregate data to deliver more holistic, integrated information to the clinician or the person being monitored. Research that has begun into this area includes the TigerPlace study, which includes bed-and-activity monitoring sensors (Rantz et al. 2012). In the study, the signals were processed locally and the processed information (i.e. events) were sent wirelessly to a central server where the information was saved in a MySQL database (Microsoft, Redmond, WA) for review by the clinician.

While only two form factors were connected in the system developed by Rantz et al. (2012), all of the form factors discussed throughout this chapter (i.e. standing, lying, sitting-form factors and activity sensors) could be used to create a network with a goal of integrating information regarding one's health. For example, each monitoring device captures an aspect of one's circadian rhythm (i.e. pattern of one's physiological condition at different times throughout the day). A network of monitoring devices would provide a more holistic picture on one's circadian rhythm, which could allow clinicians to make more well-informed assessments. In addition, a network of monitoring devices may assist in the difficult problem of monitoring the health of one or more occupants of environments that have multiple people living in them. Disambiguation could enable monitoring devices to tailor themselves to each individual. For example, following disambiguation, a monitoring device could load a configuration that is specifically set for monitoring of medical conditions of a particular individual, which would increase accuracy and overall monitoring performance. While it is inherently useful, forming a network of ambient physiological monitoring devices remains a largely unexplored area and is an interesting area in need of future work.

6.2 Clinical Evaluation of Ambient Physiological Monitoring Devices

As ambient physiological monitoring devices continue to advance, clinical aspects of the field will be examined in parallel. To be useful, ambient physiological monitoring devices need to be developed with input from target populations (including people with chronic health conditions as well as clinical professionals) followed by the evaluation of devices through structured usability studies in the real-world applications by end users.

6.2.1 Evaluation of devices on target populations

One of the important aspects of future work for ambient physiological monitoring is to appropriately match the monitoring devices with their intended target populations. For example, while a bed-form factor is currently targeted to detect problems associated with sleep disorders, it could be extended to people with other diseases, such as monitoring the vital signs of people with cardiovascular disease or tracking the nighttime wakefulness or unoccupied bed-time for people with dementia. Similarly, the scale and chair are designed mainly to aid people with heart failure while activity monitoring is intended to detect changes in activity that correspond to the progress of cognitive diseases. However, to detect and gain a better understanding of complex chronic conditions, it is likely that data from several sensor applications will have to be merged. Moreover, research thus far has conducted prototype and algorithm development and their validation using healthy adult subjects.

The authors are currently conducting trials with older adults who have a history of heart failure to simultaneously investigate the validity of novel prototype floor tile, chair and bed devices. Qualitative observations from the trials are expected to produce qualitative results. For example, older adults appeared to have more difficulty in standing without support for four to five minutes, thus creating the need for more movement artifacts compared to healthy adults.

Other studies involving the target population began to appear recently. For the bed-form factor, a number of studies compared the functionality of the device between healthy adults and people with sleep disorders (refer to Section 3.2). The weight scale and chair-form factors do not have comparison studies yet; however, Giovangrandi et al. (2012) performed the evaluation of ECG and BCG from a weight scale on people with cardiovascular disease as discussed in Section 2.4. The sitting-form factor has yet to be studied with the population with a health condition (as discussed in Section 4.3).

In terms of sensor-based activity monitoring, studies are starting to record data within real-world environments. Latest research in the field of activity recognition is looking at the real world environment with unconstrained conditions involving healthy adults (Krishnan and Cook 2014). The most recent work in the field of activity monitoring without activity recognition was successful in detecting the presence of mild cognitive impairment in older adults in real-world environments (Akl et al. 2015). Works such as these point to the trend toward studies involving people with increasingly serious health conditions in unconstrained environments.

Research on ambient physiological monitoring is slowly advancing from prototype validation with healthy adults towards validation with the population that the devices are designed for. It is anticipated that within a few years' time, more studies involving people with physiological conditions from target populations will emerge.

6.2.2 Longitudinal studies

According to Peetoom et al. (2014) and Hanson et al. (2013), physiological monitoring via environmental sensors has mainly focused on technical feasibility studies and is very limited in clinical studies, especially longitudinal studies. In agreement with this assessment, most of the ambient physiological monitoring devices discussed in this chapter have yet to be tested through longitudinal studies where devices are installed in the homes of the target populations. One research group has conducted a longitudinal study based on smart home sensors for activity monitoring and physiological monitoring. The study was conducted through TigerPlace, which is a retirement community of independent older adults designed to support aging by providing care as needed to the residents. Homes in TigerPlace have sensors installed to monitor the occupants for scientific research, residents give consent for their information to be used for this purpose (Rantz et al. 2013).

A year-long pilot study conducted at TigerPlace involved 41 older adults who used an ambient system that generated health alerts whenever an adverse event was observed, such as restlessness during sleep, which were sent to health care providers. The alert generation algorithm was developed by the retrospective analyses of 74 important health events from 16 subjects (e.g. falls, hospitalizations, emergency room visits) prior to the longitudinal study. It was observed that data from sensors such as infrared motion sensors around the house and pressure sensors on the bed began to change 10 to 14 days prior to the event; a finding that was incorporated into the algorithm (Rantz et al. 2010; Rantz et al. 2012). Over the duration of the study, 219 health events consisting of 32 emergency room visits, 23 hospitalizations and 164 falls occurred; within the last six months of the study, 258 alerts were generated by the activity sensors and 246 alerts by the bed sensor. Once an alert was received, a TigerPlace care coordinator reviewed the alert and the sensor data to determine if the alert required further assessment. Throughout the trial, the clinicians gave weekly feedback to the research group who modified the system as required. Exact accuracy of these alarms are not reported; however, it was reported that about 30 per cent of these alerts were false alarms. The clinicians involved in the study reported that this was an acceptable false alarm rate for early detection of illnesses. Based on the final 9 months of data, the clinicians rated the alerts that were most clinically relevant were recliner-chair restlessness (63 per cent), bed restlessness (75 per cent) and slow pulse in bed (100 per cent). The results also indicated that the intervention group (n = 20; subjects who were monitored) showed significant improvements compared to the control group (n = 21; subjects who were not monitored) in terms of gait assessment and right hand grip strength (Rantz et al. 2012). An interesting observation in the study was that clinicians considered the amount of activity at night (e.g. time up at night) much more clinically relevant than the increase in the activity in daytime. Many of the alerts generated by increased time up at night resulted in early detection of diseases, like urinary tract infections, pneumonia, upper respiratory infections, heart failure, pain post-hospitalization, delirium and hypoglycemia (Rantz et al. 2012). In terms of usability of the web interface for clinicians, the average time of interpretation for each alert was reduced from 4.26 minutes at the beginning of the study to 2.01 minutes by study end as the interface was revised weekly throughout the trial and the clinicians became more adept in using the interface. Finally, it is noteworthy that the TigerPlace staff adopted the alert system used for the trial after the trial was over, based on the benefits it provided during the trial period.

An issue involved with this study was that the usual interventions that took place among the participants were not controlled. The researchers pointed out that possibly the results obtained were a mixture of the ambient physiological monitoring system as well as other interventions, such as physical therapy, occupational therapy and surgery. Regardless, this study illuminates important benefits and issues regarding ambient physiological monitoring devices as they are tested in actual real-world environments. As more devices are implemented in long-term studies, it is likely that more undesirable factors (e.g. movement artifacts) that affect the accuracy and reliability of health monitoring will be revealed. These findings will guide researchers towards ways for making ambient technologies more useful in real life environments, both from the perspective of the people being monitored and as a clinical tool. In addition to the health benefits, other factors that need to be assessed through long-term studies include financial factors, clinical benefits reported by caregivers and clinicians and users' perceptions of the devices. Findings from such studies will help developers to create form factors and user interfaces that are more acceptable to the people using the devices.

In this chapter, we focused our discussion on the application of ambient physiological monitoring in a home environment; however, long-term care facilities and hospitals have just as much potential to benefit from the use of the ambient physiological monitoring as the home environment.

6.2.3 Usability studies

Users' perceptions regarding devices' usability and utility must be examined to ensure that they are appropriate and acceptable. In Section 2.4, Giovangrandi et al. (2012) reported that the weight scale had a high user acceptance due to the unobtrusiveness and intuitive use of the device. While informal user perceptions such as this are available, formal assessments of users' perceptions regarding ambient monitoring devices have not been conducted by the majority of developers. Such analyses are vital in guiding how the monitoring devices should evolve for real-life applications. A formal user perception analysis conducted on the clinicians in the TigerPlace study described above revealed shifts in overall confidence in the early illness detection system (i.e. ambient physiological monitoring system implemented at the TigerPlace) from 61 per cent to 86 per cent; ease of interpretation of the alerts and the sensor data from 43 per cent to 70 per cent; clinical relevance of the sensor data from 52 per cent to 83 per cent; confidence that the system would alert the clinicians to signs of potential decline in physical function, acute illness, or exacerbation of chronic illness from 26 per cent to 71 per cent; ease of use from 50 per cent to 85 per cent (Rantz et al. 2012). These results are promising indications that the ambient physiological monitoring devices can be designed to prove useful to people with chronic illnesses and their carers. Similarly, the authors are currently conducting qualitative analysis of older adults' perceptions regarding various ambient physiological monitoring devices.

As research on ambient physiological monitoring devices advances, more usability studies are sure to emerge, helping developers to appropriately modify devices to suit the needs of end users.

6.3 Zero-effort Technology (ZET) Revisited

This chapter has presented several different form factors that demonstrate how zero-effort technologies (ZETs) can be beneficial to people using them in many ways. ZETs, which require little or no effort from the person operating the device, are an excellent option for physiological monitoring in several ways. First of all, ZETs remove the burden of operating equipment from both the person being monitored and (if applicable) his or her caregivers. This is especially valuable for populations, such as people with dementia. As discussed in Section 1.1.1, as the severity of disease progresses, people with dementia have a greater difficulty in completing tasks and increasingly rely on help from caregivers. By using ZET, the burden of using the medical equipment is removed for both the caregiver and the person with dementia. Secondly, ZETs provide regularity of measurement. For people with conditions, such as HF, the time of the day as a physiological measurement can influence the diagnosis of the disease. As people with HF interact with zero-effort physiological monitoring devices around the house, the devices can take measurements many times a day and at any time of the day, assisting the clinicians to make more informed decision for the subject. Additionally, continuous monitoring and a more comprehensive understanding of a person's condition may help the person with the condition to take more interest and become more involved in his or her health. Another benefit of using ZETs is that energy and time that would have been used to operate conventional medical equipment can now be redirected to other activates. Lastly, zero-effort physiological monitoring may decrease the stigma of people with chronic diseases because they blend naturally into the home environment, thereby removing the necessity of having conventional, often large and obtrusive, medical devices around the home. The potential benefits of using ZETs for physiological monitoring are enormous and will surely be incorporated in future research on physiological monitoring.

7. Conclusions

In this chapter, the topic of ambient physiological monitoring was examined. An aging population, increasing number of people with chronic diseases and saturation of health-care resources are sociological driving factors for the development of ambient physiological monitoring devices. Early detection of an onset or exacerbation of a disease is part of an effective solution for chronic conditions, such as dementia and HF. Importantly, appropriate interventions can prolong independence and provide an option to age-in-place by monitoring health and enabling early interventions to improve health outcomes.

Four forms of ambient monitoring were discussed in detail—standing, bed and sitting-form factors as well as activity monitoring within an environment. The standing-form factor included the weight scale and floor-tile devices. The signals collected from the standing-form factor were BCG, weight and in some cases ECG. BCG provides heart rate and can be combined with ECG to produce a parameter called the R-J interval, which can be used to estimate the systolic blood pressure. The second form factor was the bed-form factor, which was mainly used to assess sleep disorders but can benefit people with HF and dementia as well. The primary physiological signals collected from the bed are BCG, respiration and in one case, weight. Thanks to how it is naturally used, a bed-form factor generally products significantly more clean data compared to standing or sitting-form factors. The sitting-form factor includes chairs, couches and a toilet seat. The physiological signals include BCG, PPG and ECG. CC-electrodes are being investigated as a method for collecting ECG through clothing and ECG and PPG were combined to retrieve PAT, which can be used to estimate SBP. Finally, ambient activity monitoring was discussed, where environmental sensor-based applications were used to detect a person's current location and infer details regarding the activities he or she engages in.

Based on the review of current research, it is observed that ambient physiological monitoring research is moving away from laboratory-based evaluations with healthy adults to real-world applications involving target populations. That said, there are still a number of technical improvements that must be made, such as ambient robustness to movement artifacts, disambiguation of multiple people, algorithm to detect adverse event, incorporation of textile sensors and forming a network of ambient physiological monitoring devices. From the clinical perspective, there needs to be long-term testing of ambient devices with the target populations, including usability studies conducted in real-home environments.

This chapter promotes the concept of ambient physiological monitoring in the perspective of ZETs and the devices that can achieve their intended functions without any effort by the user have a possibility to increase the usability, compliance and acceptability by the people using them. However, this notion must be validated through a long-term study of such monitoring devices in a real-home environment.

Research efforts in ambient physiological monitoring have exploded in the past decade and a plethora of devices and algorithms have emerged, enabling more effectively tailored physiological monitoring of people with chronic diseases. Physiological signals that were inaccessible or simply ignored in the past, such as BCG, are being investigated at an unprecedented rate with the help of modern electronics and computational tools. It is anticipated that this momentum will continue in the coming years and ambient physiological monitoring will likely realize its intended purpose and be available to people living in the community within the next 10 to 20 years.

Keywords: Ambient physiological monitoring, Zero-effort Technologies, Dementia, Cardiovascular Disease

References

Akhbardeh, A., S. Junnila, T. Koivistoinen and A. Varri. 2007. An intelligent ballistocardiographic chair using a novel SF-ART neural network and biorthogonal wavelets. Journal of Medical Systems. 31(1): 69–77.

Akl, A., B. Taati and A. Mihailidis. 2015. Autonomous unobtrusive detection of mild cognitive impairment in older adults. IEEE Transactions on Bio-Medical Engineering. 62(5): 1383–1394.

Alwan, M. 2009. Passive In-home health and wellness monitoring: Overview, Value and Examples. Conference Proceedings of Annual International Conference of the IEEE Engineering in Medicine and Biology Society. IEEE Engineering in Medicine and Biology Society. Conference. 2009: 4307–4310.

Alzheimer's Association. 2014. Dementia, accessed 05/25, 2015.
Alzheimer's Disease International. 2009. World Alzheimer Report 2009. London, UK: Alzheimer's Disease International.
Alzheimer's Disease International. 2014. World Alzheimer Report 2014: Dementia and Risk Reduction. London, UK: Alzheimer's Disease International.
Apple. 2015. Apple Watch User Manual: Apple.
Arcelus, A., M.H. Jones, R. Goubran and F. Knoefel. 2007. Integration of Smart Home Technologies in a Health Monitoring System for the Elderly. May.
Arcelus, Amaya, Mohammed Sardar and Alex Mihailidis. 2013. Design of a capacitive ECG sensor for unobtrusive heart rate measurements. Conference Proceedings of Instrumentation and Measurement Technology Conference (I2MTC), 2013 IEEE International.
Aubert, X.L. and A. Brauers. 2008. Estimation of vital signs in bed from a single unobtrusive mechanical sensor: algorithms and real-life evaluation. Conference Proceedings of Annual International Conference of the IEEE Engineering in Medicine and Biology Society. IEEE Engineering in Medicine and Biology Society. Conference. 2008: 4744–4747.
Austin, D., T.L. Hayes, J. Kaye, N. Mattek and M. Pavel. 2011. On the disambiguation of passively measured in-home gait velocities from multi-person smart homes. Journal of Ambient Intelligence and Smart Environments. 3(2): 165–174.
Baek, H.J., G.S. Chung, K.K. Kim, J.S. Kim and K.S. Park. 2009. Photoplethysmogram measurement without direct skin-to-sensor contact using an adaptive light source intensity control. IEEE Transactions on Information Technology in Biomedicine : A Publication of the IEEE Engineering in Medicine and Biology Society. 13(6): 1085–1088.
Baek, H.J., G.S. Chung, K.K. Kim and K.S. Park. 2012. A smart health monitoring chair for nonintrusive measurement of biological signals. IEEE Transactions on Information Technology in Biomedicine : A Publication of the IEEE Engineering in Medicine and Biology Society. 16(1): 150–158.
Baek, H.J., K.K. Kim, J.S. Kim, B. Lee and K.S. Park. 2010. Enhancing the estimation of blood pressure using pulse arrival time and two confounding factors. Physiological Measurement. 31(2): 145–157.
Bruser, C., J. Diesel, M.D.H. Zink, S. Winter, P. Schauerte and S. Leonhardt. 2013. Automatic detection of atrial fibrillation in cardiac vibration signals. IEEE Journal of Biomedical and Health Informatics. 17(1): 162–171.
Bruser, C., A. Kerekes, S. Winter and S. Leonhardt. 2012. Multi-channel optical sensor-array for measuring ballistocardiograms and respiratory activity in bed. Conference Proceedings of Annual International Conference of the IEEE Engineering in Medicine and Biology Society. IEEE Engineering in Medicine and Biology Society. Annual Conference. 2012: 5042–5045.
Bruser, C., K. Stadlthanner, A. Brauers and S. Leonhardt. 2010. Applying machine learning to detect individual heart beats in ballistocardiograms. Conference Proceedings of Annual International Conference of the IEEE Engineering in Medicine and Biology Society. IEEE Engineering in Medicine and Biology Society. Annual Conference. 2010: 1926–1929.
Bruser, C., K. Stadlthanner, S. de Waele and S. Leonhardt. 2011. Adaptive beat-to-beat heart rate estimation in ballistocardiograms. IEEE Transactions on Information Technology in Biomedicine : A Publication of the IEEE Engineering in Medicine and Biology Society. 15(5): 778–786.
Bruser, C., S. Winter and S. Leonhardt. 2013. Robust inter-beat interval estimation in cardiac vibration signals. Physiological Measurement. 34(2): 123–138.
Bruser, C., N. Strutz, S. Winter, S. Leonhardt and M. Walter. 2014. Monte-Carlo simulation and automated test bench for developing a multichannel NIR-based vital-signs monitor. IEEE Transactions on Biomedical Circuits and Systems. 9(3): 421–30.
Button, Vera. 2005. Flow transducers. *In*: Principles of Measurement and Transduction of Biomedical Variables, 305. London, UK: Academic Press.

Calhoun, D.A. and S.M. Harding. 2010. Sleep and hypertension. Chest. 138(2): 434–443.

Casanella, R., J. Gomez-Clapers and R. Pallas-Areny. 2012. On time interval measurements using BCG. Conference Proceedings of Annual International Conference of the IEEE Engineering in Medicine and Biology Society. IEEE Engineering in Medicine and Biology Society. Annual Conference. 2012: 5034–5037.

Casas, O., E.M. Spinelli and R. Pallas-Areny. 2009. Fully differential AC-coupling networks: A comparative study. IEEE Transactions on Instrumentation and Measurement. 58(1): 94–98.

Chang, H., Z. Sun, M. Saito, Q. Yuan, H. Zhang, J. Li, Z. Wang et al. 2013. Regulating infrared photoresponses in reduced graphene oxide phototransistors by defect and atomic structure control. ACS Nano. 7(7): 6310–6320.

Chaudhry, S.I., Y. Wang, J. Concato, T.M. Gill and H.M. Krumholz. 2007. Patterns of weight change preceding hospitalization for heart failure. Circulation. 116(14): 1549–1554.

Chen, L., J. Hoey, C.D. Nugent, D.J. Cook and Z. Yu. 2012. Sensor-based activity recognition. Systems, Man, and Cybernetics, Part C: IEEE Transactions on Applications and Reviews. 42(6): 790–808.

Chen, K.Y., M. Harniss, S. Patel and K. Johnson. 2014. Implementing technology-based embedded assessment in the home and community life of individuals aging with disabilities: A participatory research and development study. Disability and rehabilitation. Assistive Technology. 9(2): 112–120.

Chi, Y.M., Tzyy-Ping Jung and G. Cauwenberghs. 2010. Dry-Contact and noncontact biopotential electrodes: Methodological review. IEEE Reviews in Biomedical Engineering. 3: 106.

Dodge, H.H., N.C. Mattek, D. Austin, T.L. Hayes and J.A. Kaye. 2012. In-home walking speeds and variability trajectories associated with mild cognitive impairment. Neurology. 78(24): 1946–1952.

Domingo, Mar, Lupón, Josep, González, Beatriz, Crespo, Eva, López, Raúl, Ramos, Anna, Urrutia, Agustín, Pera, Guillem, Verdú, José M. and Bayes-Genis, Antoni. 2011. Noninvasive Remote Telemonitoring for Ambulatory Patients with Heart Failure: Effect on Number of Hospitalizations, Days in Hospital, and Quality of Life. CARME (CAtalan Remote Management Evaluation) Study. Revista Española de Cardiología (English Edition). 64(04): 277–285.

Etemadi, M., O.T. Inan, R.M. Wiard, G.T. Kovacs and L. Giovangrandi. 2009. Non-invasive assessment of cardiac contractility on a weighing scale. Conference Proceedings of Annual International Conference of the IEEE Engineering in Medicine and Biology Society. IEEE Engineering in Medicine and Biology Society. Conference. 2009: 6773–6776.

Farber, Nicholas, Douglas Shinkle, Jana Lynott, Wendy Fox-Grage and Rodney Harrell. 2011. Aging in Place: A State Survey of Livability Policies and Practices.

Friedrich, D., X.L. Aubert, H. Fuhr and A. Brauers. 2010. Heart rate estimation on a beat-to-beat basis via ballistocardiography—a hybrid approach. Conference Proceedings of Annual International Conference of the IEEE Engineering in Medicine and Biology Society. IEEE Engineering in Medicine and Biology Society. Conference. 2010: 4048–4051.

Geddes, L.A., M. Voelz, S. James and D. Reiner. 1981. Pulse arrival time as a method of obtaining systolic and diastolic blood pressure indirectly. Medical & Biological Engineering & Computing. 19(5): 671–672.

Gilaberte, S., J. Gomez-Clapers, R. Casanella and R. Pallas-Areny. 2010. Heart and respiratory rate detection on a bathroom scale based on the ballistocardiogram and the continuous wavelet transform. Conference Proceedings of Annual International Conference of the IEEE Engineering in Medicine and Biology Society. IEEE Engineering in Medicine and Biology Society. Conference. 2010: 2557–2560.

Giovangrandi, L., O.T. Inan, D. Banerjee and G. T. Kovacs. 2012. Preliminary results from BCG and ECG measurements in the heart failure clinic. Conference Proceedings of Annual

International Conference of the IEEE Engineering in Medicine and Biology Society. IEEE Engineering in Medicine and Biology Society. Annual Conference. 2012: 3780–3783.

Goldberg, L.R., J.D. Piette, M.N. Walsh, T.A. Frank, B.E. Jaski, A.L. Smith, R. Rodriguez et al. 2003. Randomized trial of a daily electronic home monitoring system in patients with advanced heart failure: the weight monitoring in heart failure (WHARF) trial. American Heart Journal. 146(4): 705–712.

Gomez-Clapers, J., R. Casanella and R. Pallas-Areny. 2012. Multi-Signal Bathroom Scale to Assess Long-Term Trends in Cardiovascular Parameters.

Gonzalez-Landaeta, R., O. Casas and R. Pallas-Areny. 2008. Heart rate detection from plantar bioimpedance measurements. IEEE Transactions on Bio-Medical Engineering. 55(3): 1163–1167.

Greenfield, Adam. 2010. Everyware: The Dawning Age of Ubiquitous Computing New Riders.

Hanson, G.J., P.Y. Takahashi and J.L. Pecina. 2013. Emerging technologies to support independent living of older adults at risk. Journal of Case Management and the Journal of Long-Term Home Health Care. 14(1): 58–64.

Heart and Stroke Foundation of Canada. Heart Failure, last modified April 20132015, http://www.heartandstroke.com/site/c.ikIQLcMWJtE/b.3484065/k.C530/Heart_disease__Heart_failure.htm.

Heart and Stroke Foundation of Canada. Statistics, last modified 20152015, http://www.heartandstroke.com/site/c.ikIQLcMWJtE/b.3483991/k.34A8/Statistics.htm.

Hong, X. and C. Nugent. 2011. Implementing evidential activity recognition in sensorised homes. Technology and Health Care: Official Journal of the European Society for Engineering and Medicine. 19(1): 37–52.

Inan, O., P.F. Migeotte, K.S. Park, M. Etemadi, K. Tavakolian, R. Casanella, J. Zanetti et al. 2014. Ballistocardiography and seismocardiography: a review of recent advances. IEEE Journal of Biomedical and Health Informatics.

Inan, O.T., M. Etemadi, A. Paloma, L. Giovangrandi and G.T. Kovacs. 2009. Non-invasive cardiac output trending during exercise recovery on a bathroom-scale-based ballistocardiograph. Physiological Measurement. 30(3): 261–274.

Inan, O.T., M. Etemadi, R.M. Wiard, G.T. Kovacs and L. Giovangrandi. 2008. Non-invasive measurement of valsalva-induced hemodynamic changes on a bathroom scale ballistocardiograph. Conference Proceedings of Annual International Conference of the IEEE Engineering in Medicine and Biology Society. IEEE Engineering in Medicine and Biology Society. Annual Conference. 2008: 674–677.

Inan, O.T., M. Etemadi, B. Widrow and G.T. Kovacs. 2010a. Adaptive cancellation of floor vibrations in standing ballistocardiogram measurements using a seismic sensor as a noise reference. IEEE Transactions on Bio-Medical Engineering. 57(3): 722–727.

Inan, O.T., G.T. Kovacs and L. Giovangrandi. 2010b. Evaluating the lower-body electromyogram signal acquired from the feet as a noise reference for standing ballistocardiogram measurements. IEEE Transactions on Information Technology in Biomedicine : A Publication of the IEEE Engineering in Medicine and Biology Society. 14(5): 1188–1196.

Inan, O.T., D. Park, G.T. Kovacs and L. Giovangrandi. 2011. Multi-signal electromechanical cardiovascular monitoring on a modified home bathroom scale. Conference Proceedings of Annual International Conference of the IEEE Engineering in Medicine and Biology Society. IEEE Engineering in Medicine and Biology Society. Conference. 2011: 2472–2475.

Jung, Da Woon, Su Hwan Hwang, Hee Nam Yoon, Y.J.G. Lee, Do-Un Jeong and Kwang Suk Park. 2014. Nocturnal awakening and sleep efficiency estimation using unobtrusively measured ballistocardiogram. IEEE Transactions on Biomedical Engineering. 61(1): 131–138.

Junnila, S., A. Akhbardeh, L.C. Barna, I. Defee and A. Varri. 2006. A Wireless Ballistocardiographic chair. Conference Proceedings of Annual International Conference of the IEEE Engineering

in Medicine and Biology Society. IEEE Engineering in Medicine and Biology Society. Annual Conference. 1: 5932–5935.
Karki, S. and J. Lekkala. 2009. A new method to measure heart rate with EMFi and PVDF materials. Journal of Medical Engineering & Technology. 33(7): 551–558.
Kato, Tsuyoshi, Akinori Ueno, Sachiyo Kataoka, Hiroshi Hoshino and Yoji Ishiyama. 2006. An Application of Capacitive Electrode for Detecting Electrocardiogram of Neonates and Infants. IEEE.
Katz, S., A.B. Ford, R.W. Moskowitz, B.A. Jackson and M.W. Jaffe. 1963. Studies of illness in the aged. the Index of Adl: A standardized measure of biological and psychosocial function. Jama. 185: 914–919.
Kaye, J.A., S.A. Maxwell, N. Mattek, T.L. Hayes, H. Dodge, M. Pavel, H.B. Jimison, K. Wild, L. Boise and T.A. Zitzelberger. 2011. Intelligent systems for assessing aging changes: home-based, unobtrusive and continuous assessment of aging. The Journals of Gerontology. Series B, Psychological Sciences and Social Sciences. 66 Suppl 1: i180–90.
Kim, J.S., Y.J. Chee, J.W. Park, J.W. Choi and K.S. Park. 2006. A new approach for non-intrusive monitoring of blood pressure on a toilet seat. Physiological Measurement. 27(2): 203–211.
Kim, J.S., K.K. Kim, H.J. Baek and K.S. Park. 2008. Effect of confounding factors on blood pressure estimation using pulse arrival time. Physiological Measurement. 29(5): 615–624.
Kim, Ko Keun, Yong Kyu Lim and Kwang Suk Park. 2006. Common Mode Noise Cancellation for Electrically Non-Contact ECG Measurement System on a Chair. IEEE.
Kortelainen, J.M., M.O. Mendez, A.M. Bianchi, M. Matteucci and S. Cerutti. 2010. Sleep staging based on signals acquired through bed sensor. IEEE Transactions on Information Technology in Biomedicine : A Publication of the IEEE Engineering in Medicine and Biology Society. 14(3): 776–785.
Krishnan, N.C. and D.J. Cook. 2014. Activity recognition on streaming sensor data. Pervasive and Mobile Computing. 10(Pt B): 138–154.
Kung, H.C., D.L. Hoyert, J. Xu and S.L. Murphy. 2008. Deaths: final data for 2005. National Vital Statistics Reports : From the Centers for Disease Control and Prevention, National Center for Health Statistics, National Vital Statistics System. 56(10): 1–120.
Lee, H., D. Lee, Y. Ahn, E.W. Lee, L.S. Park and Y. Lee. 2014. Highly efficient and low voltage silver nanowire-based OLEDs employing a N-type hole injection layer. Nanoscale. 6(15): 8565–8570.
Lim, Yong Gyu, Gih Sung Chung and Kwang Suk Park. 2009. Capacitive-driven Right-leg Grounding in Indirect-Contact ECG Measurement.
Lloyd-Jones, D., R. Adams, M. Carnethon, G. De Simone, T.B. Ferguson, K. Flegal, E. Ford et al. 2009. Heart disease and stroke statistics—2009 update: A report from the american heart association statistics committee and stroke statistics subcommittee. Circulation. 119(3): 480–486.
Mack, D.C., J.T. Patrie, P.M. Suratt, R.A. Felder and M.A. Alwan. 2009. Development and preliminary validation of heart rate and breathing rate detection using a passive, ballistocardiography-based sleep monitoring system. IEEE Transactions on Information Technology in Biomedicine: A Publication of the IEEE Engineering in Medicine and Biology Society. 13(1): 111–120.
Marek, Karen Dorman and Marilyn J. Rantz. 2000. Aging in place: a new model for long-term care. Nursing Administration Quarterly. 24(3): 1–11.
Mercado, Ruben, A. Paez, D.M. Scott, K.B. Newbold and P. Kanaroglou. 2007. Transport policy in aging societies: an international comparison and implications for canada. Transport Policy. 1: 1.
Migliorini, M., A.M. Bianchi, D. Nistico, J. Kortelainen, E. Arce-Santana, S. Cerutti and M.O. Mendez. 2010. Automatic sleep staging based on ballistocardiographic signals recorded through bed sensors. Conference Proceedings of Annual International Conference of the

IEEE Engineering in Medicine and Biology Society. IEEE Engineering in Medicine and Biology Society. Annual Conference. 2010: 3273–3276.

Mihailidis, Alex, Jennifer Boger, Jesse Hoey and Tizneem Jiancaro. 2011. Zero effort technologies: considerations, challenges and use in health, wellness and rehabilitation. Synthesis Lectures on Assistive, Rehabilitative and Health-Preserving Technologies. 1(2): 1–94.

Muehlsteff, J., T. Correia, R. Couceiro, P. Carvalho, A. Ritz, C. Eickholt, M. Kelm and C. Meyer. 2013. Detection of hemodynamic adaptations during impending syncope: implementation of a robust algorithm based on pulse arrival time measurements only. Conference Proceedings of Annual International Conference of the IEEE Engineering in Medicine and Biology Society. IEEE Engineering in Medicine and Biology Society. Annual Conference. 2013: 2291–2294.

Murphy, Kevin P. 2012. Machine Learning: A Probabilistic Perspective. MIT Press.

Odinaka, I., Po-Hsiang Lai, A.D. Kaplan, J.A. O'Sullivan, E.J. Sirevaag and J.W. Rohrbaugh. 2012. ECG biometric recognition: a comparative analysis. IEEE Transactions on Information Forensics and Security. 7(6): 1812–1824.

Ory, M.G., R.R. Hoffman 3rd, J.L. Yee, S. Tennstedt and R. Schulz. 1999. Prevalence and impact of caregiving: a detailed comparison between dementia and nondementia caregivers. The Gerontologist. 39(2): 177–185.

Owan, T.E. and M.M. Redfield. 2005. Epidemiology of diastolic heart failure. Progress in Cardiovascular Diseases. 47(5): 320–332.

Paalasmaa, J., M. Waris, H. Toivonen, L. Leppakorpi and M. Partinen. 2012. Unobtrusive online monitoring of sleep at home. Conference Proceedings of Annual International Conference of the IEEE Engineering in Medicine and Biology Society. IEEE Engineering in Medicine and Biology Society. Annual Conference. 2012: 3784–3788.

Peetoom, K.K., M.A. Lexis, M. Joore, C.D. Dirksen and L.P. De Witte. 2014. Literature review on monitoring technologies and their outcomes in independently living elderly people. Disability and Rehabilitation. Assistive Technology. 1–24.

Public Health Agency of Canada. 2014. The Chief Public Health Officer's Report on the State of Public Health in Canada, Public Health in the Future—Changing Demographics, Aging and Health, last modified 2014-09-08, accessed 05/25, 2015, http://www.phac-aspc.gc.ca/cphorsphc-respcacsp/2014/chang-eng.php.

Rantz, M.J., M. Skubic, G. Alexander, M.A. Aud, B.J. Wakefield, C. Galambos, R.J. Koopman and S.J. Miller. 2010. Improving nurse care coordination with technology. Computers, Informatics, Nursing : CIN. 28(6): 325–332.

Rantz, M.J., M. Skubic, R.J. Koopman, G.L. Alexander, L. Phillips, K. Musterman, J. Back et al. 2012. Automated technology to speed recognition of signs of illness in older adults. Journal of Gerontological Nursing. 38(4): 18–23.

Rantz, M.J., M. Skubic, S.J. Miller, C. Galambos, G. Alexander, J. Keller and M. Popescu. 2013. Sensor technology to support aging in place. Journal of the American Medical Directors Association. 14(6): 386–391.

Rassaf, T., J. Muehlsteff, O. Such, M. Kelm and C. Meyer. 2010. The pulse arrival time approach for non-invasive hemodynamic monitoring in low-acuity settings. Medical Science Monitor : International Medical Journal of Experimental and Clinical Research. 16(11): MT83–7.

Reisberg, Barry, Sanford Finkel, John Overall, Norbert Schmidt-Gollas, Siegfried Kanowski, Hartmut Lehfeld, Franz Hulla, Steven G. Sclan, Hans-Ulrich Wilms and Kurt Heininger. 2001. The Alzheimer's Disease Activities of Daily Living International Scale (ADL-IS). International Psychogeriatrics. 13(02): 163–181.

Remme, W.J. and K. Swedberg. 2001. Guidelines for the diagnosis and treatment of chronic heart failure. European Heart Journal. 22(17): 1527–1560.

Rosales, L., M. Skubic, D. Heise, M.J. Devaney and M. Schaumburg. 2012. Heartbeat detection from a hydraulic bed sensor using a clustering approach. Conference Proceedings of Annual International Conference of the IEEE Engineering in Medicine and Biology Society. IEEE Engineering in Medicine and Biology Society. Annual Conference. 2012: 2383–2387.

Samsung. 2015. Samsung Gear S User Manual. Samsung.

Shin, J.H., B.H. Choi, Y.G. Lim, D.U. Jeong and K.S. Park. 2008. Automatic ballistocardiogram (BCG) beat detection using a template matching approach. Conference Proceedings of Annual International Conference of the IEEE Engineering in Medicine and Biology Society. IEEE Engineering in Medicine and Biology Society. Conference. 2008: 1144–1146.

Shin, J.H., K.M. Lee and K.S. Park. 2009. Non-Constrained Monitoring of Systolic Blood Pressure on a Weighing Scale. Physiological Measurement. 30(7): 679–693.

Sola, J., R. Vetter, P. Renevey, O. Chetelat, C. Sartori and S.F. Rimoldi. 2009. Parametric Estimation of Pulse Arrival Time: A Robust Approach to Pulse Wave Velocity. Physiological Measurement. 30(7): 603–615.

Starr, Isaac, A.J. Rawson, H.A. Schroeder and N.R. Joseph. 1939. Studies on the estimation of cardiac ouptut in man, and of abnormalities in cardiac function, from the heart's recoil and the blood's impacts; the ballistocardiogram. American Journal of Physiology—Legacy Content. 127(1): 1–28.

Stone, E.E. and M. Skubic. 2011. Passive in-Home Measurement of Stride-to-Stride Gait Variability Comparing Vision and Kinect Sensing. Aug.

Su, B.Y., K.C. Ho, M. Skubic and L. Rosales. 2012. Pulse rate estimation using hydraulic bed sensor. Conference Proceedings of Annual International Conference of the IEEE Engineering in Medicine and Biology Society. IEEE Engineering in Medicine and Biology Society. Annual Conference. 2012: 2587–2590.

University of Ottawa Heart Institute. 2015. Heart Failure: A Guide for Patients and Families, last modified May 25, 2015, http://www.ottawaheart.ca/content_documents/HeartFailure-Guide-UOHI01-Eng-Web.pdf.

Waite, L.M., D.A. Grayson, O. Piguet, H. Creasey, H.P. Bennett and G.A. Broe. 2005. Gait slowing as a predictor of incident dementia: 6-Year longitudinal data from the sydney older persons study. Journal of the Neurological Sciences. 230: 89–93.

Watanabe, K., T. Watanabe, H. Watanabe, H. Ando, T. Ishikawa and K. Kobayashi. 2005. Noninvasive measurement of heartbeat, respiration, snoring and body movements of a subject in bed via a pneumatic method. IEEE Transactions on Bio-Medical Engineering. 52(12): 2100–2107.

Watanabe, T. and K. Watanabe. 2004. Noncontact method for sleep stage estimation. IEEE Transactions on Bio-Medical Engineering. 51(10): 1735–1748.

Wiard, R.M., O.T. Inan, B. Argyres, M. Etemadi, G.T. Kovacs and L. Giovangrandi. 2011. Automatic detection of motion artifacts in the ballistocardiogram measured on a modified bathroom scale. Medical & Biological Engineering & Computing. 49(2): 213–220.

Willemen, T., D. Van Deun, V. Verhaert, S. Van Huffel, B. Haex and J. Vander Sloten. 2014. Characterization of the respiratory and heart beat signal from an air pressure-based ballistocardiographic setup. Conference Proceedings of Annual International Conference of the IEEE Engineering in Medicine and Biology Society. IEEE Engineering in Medicine and Biology Society. Annual Conference. 2014: 6298–6301.

Wong, M.Y., E. Pickwell-MacPherson, Y.T. Zhang and J.C. Cheng. 2011. The effects of pre-ejection period on post-exercise systolic blood pressure estimation using the pulse arrival time technique. European Journal of Applied Physiology. 111(1): 135–144.

Wu, Kin-fai and Yuan-ting Zhang. 2008. Contactless and Continuous Monitoring of Heart Electric Activities through Clothes on a Sleeping Bed. May.

Yamada, T., Y. Hayamizu, Y. Yamamoto, Y. Yomogida, A. Izadi-Najafabadi, D.N. Futaba and K. Hata. 2011. A stretchable carbon nanotube strain sensor for human-motion detection. Nature Nanotechnology. 6(5): 296–301.

Yao, Y., C. Bruser, U. Pietrzyk, S. Leonhardt, S. van Waasen and M. Schiek. 2014. Model-based verification of a non-linear separation scheme for ballistocardiography. IEEE Journal of Biomedical and Health Informatics. 18(1): 174–182.

Zhang, G., M. Gao, D. Xu, N.B. Olivier and R. Mukkamala. 2011a. Pulse arrival time is not an adequate surrogate for pulse transit time as a marker of blood pressure. Journal of Applied Physiology (Bethesda, Md.: 1985). 111(6): 1681–1686.

Zhang, G., D. Xu, N.B. Olivier and R. Mukkamala. 2011b. Pulse arrival time is not an adequate surrogate for pulse transit time in terms of tracking diastolic pressure. Conference Proceedings of Annual International Conference of the IEEE Engineering in Medicine and Biology Society. IEEE Engineering in Medicine and Biology Society. Annual Conference. 2011: 6462–6464.

Zheng, Y.L., X.R. Ding, C.C. Poon, B.P. Lo, H. Zhang, X.L. Zhou, G.Z. Yang, N. Zhao and Y.T. Zhang. 2014. Unobtrusive sensing and wearable devices for health informatics. IEEE Transactions on Bio-Medical Engineering. 61(5): 1538–1554.

4

Designing Formally-controlled Smart Home Systems for People with Disabilities

Sébastien Guillet, Bruno Bouchard* and *Abdenour Bouzouane*

1. Introduction

Ubiquitous computing, making us more connected to our environment and other people, is challenging the way we live through different means, ranging from anticipating our needs to securing our environment and automating routine physical tasks. Contributions to ubiquitous computing has led the scientific community to the smart home era (Ramos et al. 2008), which involves a wide range of means to liberate us from usually hard and repetitive work at home and live more independently.

Enhancing independence is actually the core concept of smart homes dedicated to disabled people. For example, such a house can be designed to help a human resident suffering from a cognitive deficit to complete his activities of daily living (ADL) (Carberry 2001) without the need of

LIARA, Université du Québec À Chicoutimi. 555 blvd Université EST, G7H2B1, Chicoutimi, Québec, Canada.
Email: {sebastien.guillet1; bruno.bouchard; abdenour.bouzouane}@uqac.ca http://liara.uqac.ca
* Corresponding author

additional human assistance. Designing such smart homes involves many challenges, including blending unobtrusively into the home environment (Novak et al. 2012), recognizing the ongoing inhabitant activity (Bouchard et al. 2007), localizing objects (Fortin-Simard et al. 2012), adapting assistance to the person's cognitive deficit (Lapointe et al. 2012) and securing the environment (Pigot et al. 2003).

Given the high degree of vulnerability of people with cognitive deficiencies, securing the house is a primary concern. Indeed, an adequately designed smart home for disabled people should be able to provide both assistance and protection. However, even if a smart home system is usually build to last, it might no be the case for its very own components (Bulow 1986): lights, screens, sound system and many other important equipments can fail during the lifetime of the system. In this context, providing a viable security strategy over time requires taking into account failures due to their high probability and potential harmful consequences, if not taken seriously (Chetan et al. 2005).

To operate properly over time, the main concern of a fault-tolerant smart home system is, upon detection of a failure, knowing how to 'react appropriately'. Let's suppose that a detected random failure affects an arbitrary component, is the system still able to provide both protection and adequate assistance with respect to the person's disabilities?

Answering such a question usually requires to solve a non-trivial combinatorial problem: a smart home is supposed to be composed of many dynamical components (electrical shutters, lights, ventilation systems, etc.), each one having several exclusive execution modes (opening, opened, on, off, disabled, failed, etc.) which can be observed, using sensors. These components are concurrently executed and their execution modes can be influenced upon reception of events which can be external to the system (e.g. the user pushes a button) and/or internal (e.g. a security system prevents a hair dryer from powering on because it is too hot). Here lies the complexity—ensuring that the system will respect a security property (e.g. being able to provide assistance even if a component fails) requires to verify that this property holds for each accessible combination of execution modes.

Now let's introduce the notion of controllability, which happens when a component offers an interface so that a control system—a program named controller—can send events to constrain its behavior. Controllability is very common in the context of ubiquitous computing—smart home is no exception—where almost every component provides such an interface so that it can be adapted to a situation. A system is said to be controllable with respect to a temporal property; whether given its dynamicity and controllability, it has a controller which constrains the system so that the temporal property holds for all possible executions.

Applied to smart homes for disabled people, a smart home has the capacity to undertake a security constraint (or, to be more general, a 'quality of service constraint') over time if and only if, a control system can be proven to keep the system in execution mode, complying with the constraint. But even if a system under control can be proven correct using verification techniques, its controller is not guaranteed to be interesting. For example, let's take a controllable component which can be prevented to start and a security constraint such that this component must not be started when temperature is above 50°C. Now let's build a controller which always disables this component; then the system under control can be verified to be correct. However, we understand that the implemented controller should be more permissive when the temperature is lower than 50°C.

Basically, in the presence of controllability, the designer of a fault-tolerant smart home system has to face two non-trivial problems: building a permissive controller and verifying the system under control. It happens that these two problems are the specialty of a formal technique named Discrete Controller Synthesis (DCS) (Ramadge and Wonham 1989): given a system's dynamicity, controllability and temporal constraints, if a control solution exists then DCS is able to provide automatically a controller which is both correct by construction and maximally permissive, meaning that it valuates control events tied to control interfaces of dynamic components only when the system has to be constrained.

In (Guillet et al. 2012), we made a contribution, giving concepts on representing the behavior of a smart home system dedicated to disabled people. DCS was shown to be applicable using this representation to create a controller designed to keep the smart home in a correct state. The contribution of the present study relies on the definition of a design methodology around these concepts and shows examples on how a smart home system can be specified, so that DCS can be applied to solve concrete fault tolerance-related problems in smart homes for people with impairments.

The paper is organized as follows: Section 2 presents the related work on security in the smart home domain and justifies the choice of DCS over other formal techniques to provide a solution for the controllability problem. Section 3 describes the synchronous framework which serves as a foundation for modeling and applying DCS. Section 4 explains how to define a smart home model using this synchronous framework. Section 5 shows the application of DCS on such a model. Section 6 details the experiments we conducted, using a partial model of our own smart home equipment. The model is kept partial so that both DCS application and controller execution remain easy to follow step by step. Finally, Section 7 concludes the paper and outlines future developments in this field.

2. Related Work

The literature on which this study is based can be divided into three majors domains. The first one is smart home modeling (Pigot et al. 2003a; Pigot et al. 2003b; Latfi et al. 2007), which aims to give a framework to represent key aspects of a smart home (dynamicity, controllability and temporal constraints) so that formal techniques, such as DCS, can be applied. These aspects are generic and need concrete definitions in the smart home context. The second domain is smart home security (Chetan et al. 2005; Lapointe et al. 2012; Augusto and Nugent 2006; Patterson et al. 2006; Mihailidis et al. 2007; Bouchard et al. 2012; Bouchard et al. 2013; Carberry 2001)—what makes a smart home secure, especially a smart home for disabled people? What are the techniques employed to provide some form of security in this context? Finally, the third domain is formal techniques (Benveniste et al. 2003; Marchand et al. 2000; Dumitrescu et al. 2010; Delaval et al. 2010; Guillet et al. 2012)—failing to provide a correct smart home behavior for all its possible executions could lead to harmful consequences for a vulnerable inhabitant; so how do we prove a smart home to be secure? And what is the best technique to apply, given the smart home properties?

2.1 Smart Home Modeling

Research projects related to modeling of smart homes for disabled people usually share many concepts based on representing: the smart home elements (devices, doors, lights, etc.) with their positions and execution modes, the person itself (its state of mind, behavior, position, cognitive profile, etc.) and the global execution model (how the smart home is supposed to run and process events in order to provide both assistance and security through artificial intelligence).

In (Pigot et al. 2003a) and (Pigot et al. 2003b), Pigot et al. present respectively (1) a metamodel containing generic knowledge of a smart home system for elders suffering from dementia, and (2) a corresponding model showing cognitive assistance and telemonitoring concepts. These works detail a pervasive infrastructure and applications to provide assistance to elders with cognitive deficiencies, using two kinds of interventions—one operating inside the home to help the person to complete its ADL in case of difficulties and another in establishing communication outside the home to send message to caregivers, medical teams or families.

In (Latfi et al. 2007), Latfi et al. give an overview of an ontology-based model of a smart home dedicated to the elderly suffering from loss of cognitive autonomy. The ontological architecture is partitioned into seven sub-domains: (1) habitat, describing the home structure (rooms, doors, windows, etc.); (2) person, which can describe the patient itself (medical history, behavior, etc.) and the various persons supposed to interact with the patient and/or the habitat (medical actor, habitat-staff, friend, etc.); (3) equipment, which defines the various home appliances; (4) software, describing reusable software modules of the smart home system; (5) task, detailing the observable tasks that the patient, the personal and house itself, can perform; (6) behavior, regrouping life habits and critical physiological parameters; and (7) decision, related to the smart home adaptation behavior.

These works constitute the foundation of Section 4, which will synthesize and show how to represent their ideas in a formal synchronous model, so that the security properties can be set and verified.

2.2 Smart Home Security

Due to its importance, security in the context of smart homes, especially those dedicated to disabled people, has been widely covered in literature. Methods employed to secure a smart home target three main layers: (1) fault tolerance, as a smart home system is supposed to experience failures through its lifetime; (2) smart sensing technology, so that ADL can be monitored accurately; and (3) appropriate smart home behavior depending on the patient deficiencies.

Fault tolerance

Failures in a smart home system may occur on several levels (Kilgore et al. 2004; Chetan et al. 2005)—a smart home is typically a set of hardware and software components communicating together, so that failures can happen either at hardware, software or communication level. Sensors, actuators, displays, speakers, lights, etc. are traditional failure-prone smart-home hardware components. They wear out over time, can be damaged, can go down if they are battery powered, can cease to communicate because of limited signal strength, can operate incorrectly because of a manufacturing defect, etc. A single failure at this level can compromise the smart home security.

A smart home system also typically contains multiple software components running together (operating systems, artificial intelligences, controllers, etc.), including commercial applications (i.e. trusted black boxes). Unless formally checked against security requirements, a few assumptions should be made about applications. Even software verified by competent and credible experts can contain bugs. The malfunction in the control software in Ariane 5 Flight 501 is an example of such a bug, which remained undetected through several human-driven verification processes (Le Lann 1996).

Communication between hardware and software components happens through wired and wireless channels. Communication failures are mainly caused by low signal strength (e.g. two mobile wireless devices communicating together get separated by too long a distance) or heavy traffic. They are not really hardware or software related, but can be (wrongly) perceived as such because affected components cease to communicate and become unavailable, making these failures important to detect.

When a hardware, software or communication component failure is detected, two common responses are: (1) using an equivalent component (redundancy) (Bouchard et al. 2012) and (2) executing the system in a degraded mode, allowing it to work correctly through failures using a safe subset of its functionalities (Jaygarl et al. 2008).

These methods will be used in Sections 4 and 5 as a base to build a fault-tolerant smart home.

Smart sensing

Increasing a smart home robustness also involves an effective sensing system. Identifying ADL (Patterson et al. 2006) (Bouchard et al. 2007), locating a person or mobile components (Fortin-Simard et al. 2012), recognizing the mood (Picard 1995), etc. are examples of smart sensing features that can be integrated into a smart home. Section 4 takes the presence of these kinds of high-level sensors (artificial intelligence) into account, so that security rules can be based on their information. In (Bouchard et al. 2012), Bouchard et al. give guidelines to integrate and execute artificial intelligence modules into a generic smart home system. We will take advantage of these guidelines to model a system that will comply with the same execution principles.

Impairment adaptation

Knowing how to adapt a smart home for disabled people to their impairments is a sensitive and complex problem largely discussed in relevant literature. Smart homes usually contain technological devices aimed to provide adapted cognitive assistance, or prompts, when needed. Typical prompts can be based on sound, music, spoken messages, photos, videos, lights, etc. Implementing an adequate prompting system is actually the core concept or impairment adaptation (Pigot et al. 2003; Patterson et al. 2006; Mihailidis et al. 2007; Lapointe et al. 2012).

In (Bouchard et al. 2012) and (Lapointe et al. 2012), the authors provide experimental results on prompt efficiency according to cognitive profiles. Section 4 shows how to represent these relations, so that security properties can be defined for the prompting system. Combined together, all these security layers raise a new question. If the house is equipped with redundant critical equipment, if the prompting system can be adapted in accordance with the severity and characteristics of the patient's impairment, and if the context (ADL, mood, position, etc.) can be accurately monitored, how do we prove that, in case of a failure, the smart home system can still provide adequate assistance if the failure impacts the prompting system or the way ADL can be monitored? Proving it for every allowed failure, every possible execution, every context, etc. is essentially a combinatorial explosion problem that is very difficult to solve without appropriate tools. This is where verification techniques come in.

2.3 Formal Methods

Many research work contribute to formal modeling and verification of users' interactions, hardware/software components and control algorithms in the smart home domain (Schmidtke and Woo 2009; Corno and Sanaullah 2011; Corno and Sanaullah 2013). However, formal verification supposes that a complete system can be modeled before being applied. In the modeling methodology proposed in (Corno and Sanaullah 2013), a modeling step named 'control algorithm modeling' is explicitly required. This step is about the definition of a module which, given (1) the system current configuration, (2) incoming message from the system or its environment, and (3) control rules, makes a reconfiguration decision and sends triggering messages to the associated devices for performing the required operations. This step is precisely the part that is difficult to design because of the combinatorial problem we are facing in this context. This is the reason why we are more interested in an alternative method, DCS, which is able to both build the control part automatically and perform formal verification of the system.

Regarding smart homes, a smart home system can be considered as a specialization of autonomic computing systems (Kephart and Chess 2003), which adapt and reconfigure themselves through the presence of a feedback loop. This loop takes inputs from the environment (e.g. sensors), updates a representation (e.g. Petri nets, automata) of the system under control and decides to reconfigure the system if necessary. This consideration is detailed in Section 3. Describing such a feedback loop can be done in terms of a DCS problem. It consists of considering on the one hand, the set of possible behaviors of a discrete event system (Cassandras and Lafortune 2006), where variables are partitioned into uncontrollable and controllable ones. The uncontrollable variables typically come from the system's environment (i.e. 'inputs'), while the values of the controllable variables are given by the synthesized controller itself. On the other hand, it requires a specification of a control objective—a property typically concerning reachability or invariance of a state space subset. Such a programming makes use of reconfiguration policy by logical contract; viz. specifications with contracts amount to specify the control objective and to have an automaton describing possible behaviors rather than writing down the complete correct control solution. The basic case is that of contracts on logical properties, i.e. involving only boolean conditions of states and events. Within the synchronous approach (Benveniste et al. 2003), DCS has been defined and implemented as a tool integrated with synchronous languages (Sigali (Marchand et al. 2000)): It handles transition systems with multi-event labels, typical of the synchronous approach and features weight functions mechanisms to introduce some quantitative information and perform optimal DCS. One of the synchronous languages it has been integrated with is BZR (Delaval et al. 2010), which is used in this work; BZR actually includes a DCS usage from Sigali within its compilation. The compilation yields (if it exists) the code of a correct-by-construction controller (here in C language), which can itself be compiled to be executed into the smart home system.

Based on the synchronous characteristics of a smart home system, Section 3 sets the synchronous context and notations so that they can be applied to smart home modeling in order to perform DCS. DCS has already been successfully applied in various domains, e.g. adaptive resource management (Delaval and Rutten 2010), reconfigurable component-based systems (Bouhadiba et al. 2011), reconfigurable embedded systems (Guillet et al. 2012), etc. However, DCS in the context of smart homes has not been seen until very recently, where it was introduced in (Zhao et al. 2013) and (Guillet et al. 2013). Both studies show preliminary results on how DCS could be applied to secure a smart home (Zhao et al. 2013), having a general point of view, and (Guillet et al. 2013) a specific one regarding fault tolerance. They have a common perspective to show results with more objectives and adaptive control in order to go beyond a demonstration of DCS applicability and really show its relevance and efficiency in this context. This study takes this perspective into account to give a contribution to actual usage of DCS to solve concrete smart home problems—related to fault tolerance—through a modeling methodology, using BZR.

3. Synchronous Framework: Basic Notions

Synchronous languages are optimized for programming reactive systems, i.e. systems that react to external events. This section aims at presenting the similarities between a reactive system under control and a controlled smart home, so that a synchronous framework, essentially adopted from (Marchand and Samaan 2000), gets justified as appropriate to specify smart home systems.

3.1 Execution Model

In (Guillet et al. 2012), the execution model of a reactive system under control is depicted in Fig. 1. Such a system contains a global execution loop, which starts by taking events from the environment. Then these events get processed by a task (reconfiguration controller), which chooses the system's configuration. Finally, this configuration order gets dispatched through the system's tasks following its model of computation and another iteration of the loop can start again. If a system can be represented within this execution model, then the proposition of this work can help to design and formally obtain its reconfiguration controller task. In (Bouchard et al. 2012), guidelines to build the software architecture of a smart home system are presented in Fig. 2. Such software follows a loop-based execution, in which a database containing an updated system state and event values is read and processed by eventual artificial intelligence (AI) modules to transform raw data into high-level information. This information can then be used through third party applications.

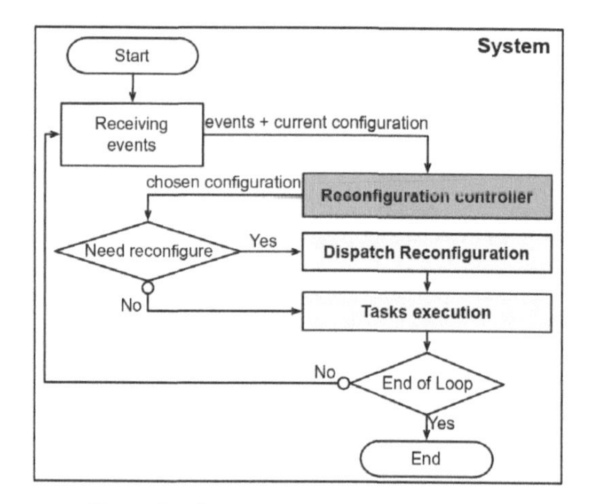

Fig. 1. Configuration processing flowchart.

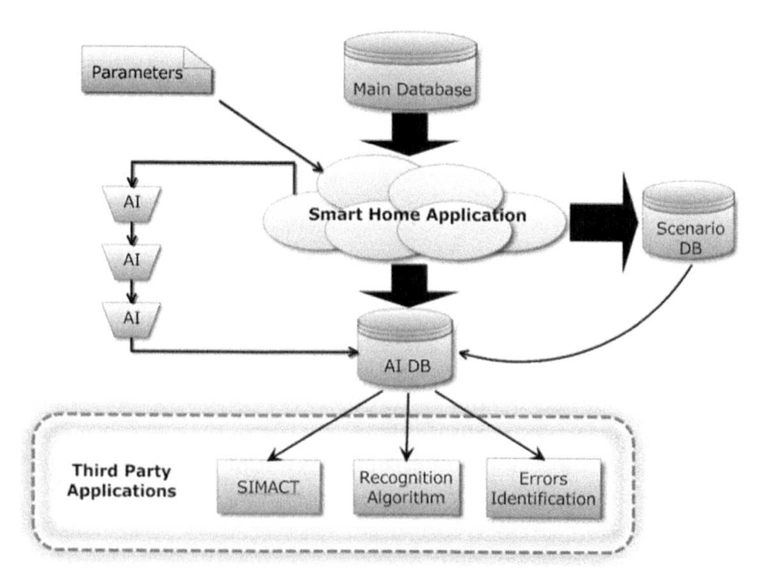

Fig. 2. Smart home software architecture.

Immediately, we can see similarities arising from such an architecture compared to the reactive system execution model. If we add a reconfiguration controller as a third party application in this software architecture, then we obtain the same execution principle as presented in Fig. 1. In each iteration of the execution loop a controller can be designed to (1) take events and/

or high level information provided by the system and its environment, (2) perform a reconfiguration decision, and (3) give this decision back to the system, using some of its actuators (i.e. its controllability) before the next iteration.

Designing the aforementioned controller by constraint so that it can be obtained automatically through DCS becomes possible, but it requires the use of a formal model to specify the behavior of the underlying system under control. Behavioral modeling can be performed using various formal representations, e.g. Statecharts, Petri-nets, Communicating Sequential Processes or other ways. The toolset we use in this work, BZR and Sigali, brings us to define our system in terms of synchronous equations and Labelled Transition Systems.

3.2 Synchronous Equation

In a declarative synchronous language, semantic is expressed in terms of dataflows—values carried in discrete time are considered as infinite sequence of values, or flows. At each discrete instant, the relation between input and output values is defined by an equational representation between flows. It is basically a system of equations, which are evaluated concurrently in the same instant and not in sequence, the real evaluation order being determined at compile-time from their interdependencies. For example, let x and y be two dataflows such that $x = x0, x1, \dots$ and $y = y0, y1, \dots$. Evolution of y over time is given by the following system of equations:

$$\begin{cases} y0 = x0 \\ yt = yt{-}1 + xt \ \mathit{if} \ t \geq 1 \end{cases}$$

In this example, y is defined, amongst others, by a reference to its value at a previous discrete instant. Each declarative synchronous language has a syntax to define such a system. The corresponding BZR program is: y = x –> pre(y) + x; meaning that in the first step, y takes the current value of x, and for all next steps y will take its previous value incremented by x. (Other syntactic features of BZR can be found online at http://bzr.inria.fr/pub/bzr-manual.pdf.) To represent the system execution modes, BZR also defines automata, or Labelled Transition Systems, with each state encapsulating a set of synchronous equations evaluated only when the state is activated.

3.3 Labelled Transition System (LTS)

LTS is a structure S = ⟨Q, q0, I, O, T⟩ where Q is a finite set of states, q0 is the initial state of S, I is a finite set of input events (produced by the environment), O is a finite set of output events (emitted towards the environment) and T is the transition relation, that is a subset of Q×Bool(I) × O* × Q, where Bool(I) is the set of boolean expressions of I. If we denote by B the set {true, false}, then a guard g ∈ Bool(I) can be equivalently seen as a function from 2^I into B.

Each transition has a label of the form *g/a,* where *g* ∈ Bool(I) must be true for the transition to be taken (*g* is the guard of the transition), where *a* ∈ O* is a conjunction of outputs that are emitted when the transition is taken (*a* is the action of the transition). State *q* is the source of the transition Tr(q, *g, a, q'*), and state *q'* is the destination.

The composition operator of two LTS put in parallel is the synchronous product, noted ||, and a characteristic feature of the synchronous languages. The synchronous product is commutative and associative. Formally: ⟨Q1, q0, 1, I1, O1, T1⟩ || ⟨Q2, q0, 2, I2, O2, T2⟩ = (g1 ∧g2)/(a1 ∧a2) ⟨Q1 × Q2, (q0, 1, q0, 2), I1 ∪ I2, O1 ∪ O2, T ⟩ with T = {Tr((q1, q2), (g1∧g2), (a1∧a2), (q'1, q'2)) | Tr(q1, g1, a1, q'1)∈ T1, Tr(q2, g2, a2, q'2) ∈ T2}.

Note that this synchronous composition is the simplified one presented in and supposes that *g* and *a* do not share any variable, which would be permitted in synchronous languages like Esterel.

Here (*q*1, *q*2) is called a macro-state, where *q*1 and *q*2 are its two component states. A macro-state containing one component state for every LTS synchronously composed in a system *S* is called a configuration of *S*.

3.4 Discrete Controller Synthesis (DCS) on LTS

A system *S* is specified as *LTS*, more precisely as the result of the synchronous composition of several *LTS*. *F* is the objective that the controlled system must fulfil and H is the behavior hypothesis on the inputs of *S*. The controller *C* obtained with *DCS* achieves this objective by restraining the transitions of *S*, that is, by disabling those that would jeopardize the objective *F*, considering the hypothesis *H*. Both *F* and *H* are expressed as boolean equations. The set I of inputs of *S* is partitioned into two subsets: the set *IC* of controllable variables and the set *IU* of uncontrollable inputs. Formally, *I* = *IC* ∪ *IU* and *IC* ∩ *IU* = ∅. As a consequence, a transition guard *g* ∈ Bool(*IC* ∪ *IU*) can be seen as a function from *2IC* × *2IU* into *B*.

A transition is controllable if and only if there exists at least one valuation of the controllable variables such that the boolean expression of its guard is false; otherwise, it is uncontrollable. Formally, a transition $Tr(q, g, a, q')$ $\in T$ is controllable if $\exists X \in 2IC$ such that $\forall Y \in 2IU$, we have $g(X,Y)$ = false. In the proposed framework, the following function Sc = make invariant (S,E) from Sigali is used to synthesize (i.e. compute by inference) the controlled system $Sc = S \parallel C$ where E is any subset of states of S, possibly specified itself as a predicate on states (or control objective) F and predicate on inputs (or hypothesis) H. The function makes invariant synthesizes and returns a controllable system Sc, if it exists, such that the controllable transitions leading to states $qi \notin E$ are inhibited, as well as those leading to states from where a sequence of uncontrollable transitions can lead to such states $qi \notin E$. If *DCS* fails, it means that a controller of S does not exist for objective F and hypothesis H. In this context, the present proposition relies on the use of *DCS* to synthesize a controller C, which makes invariant a safe set of states E in a LTS-based system where E is inferred by boolean equations defining a control objective and an hypothesis on the inputs. The controller C given by *DCS* is said to be maximally permissive, meaning that it doesn't set values of controllable variables that can either be true or false while still compliant with the control objective. Actually, the BZR compiler defaults these variables as true. Optimization can be done at this level if this type of decision is too arbitrary (Guillet et al. 2012), but it goes beyond the scope of this work, which focuses on security. So the standard decision behavior given by BZR is kept. A smart home system, following the aforementioned execution principle, can now be designed using this framework.

4. Smart Home Model

From the various smart home presentations found in the related work, a smart home system for people with disabilities can be abstracted as a hierarchy of hardware and software components (dynamic or not), sensors and effectors distributed among several interconnected rooms, helping a person with impairments to perform ADL. Showing how to specify all these features within a synchronous model is the aim of this section.

4.1 Dynamic Components

The top component of the hierarchy is the system itself. In accordance with the synchronous execution model, let S be the LTS of the system, taking inputs I from its effectors (buttons, touchscreens, controllable interfaces, etc.) and producing outputs *O* from its sensors (low level sensors, *AI*, any device producing notifications, etc.) each time it is triggered.

The smart home system is usually built upon several components, which can in turn be defined as LTS or LTS compositions if they are dynamic (i.e. they have multiple exclusive running modes) or a set of synchronous equations if they have only one execution mode. Some components may be redundant and should not be specified more than once. For this case, BZR provides a node construct, in which LTS and synchronous equations can be defined to be instantiated. Figure 3 shows the graphical representation of such a node for a light bulb behavior definition.

Representing or not a component must be decided upon the following principle: if a component is concerned by a security rule, or if it can directly or indirectly influence a component concerned by a security rule, its behavior must be defined in the synchronous model. Moreover, if a behavior is modeled, it must also be observable. Regarding the example of the light bulb from Fig. 3, if its corresponding switch is set to ON or OFF (the state of the switch is itself supposed to be known by the system), then the bulb is supposed to respectively light up or shut down. This abstraction can work for a system with a relatively short life and built with new light bulbs. However, in the context of smart homes, a light bulb may fail at some point. In this model, the light bulb failure is not observable, so it does not correspond to reality. Being able to observe such a failure requires another component, like an appropriate sensor represented in Fig. 4 by the boolean variable lightIsOn. To keep track of the failure, it can be represented as an execution mode, as in Fig. 5.

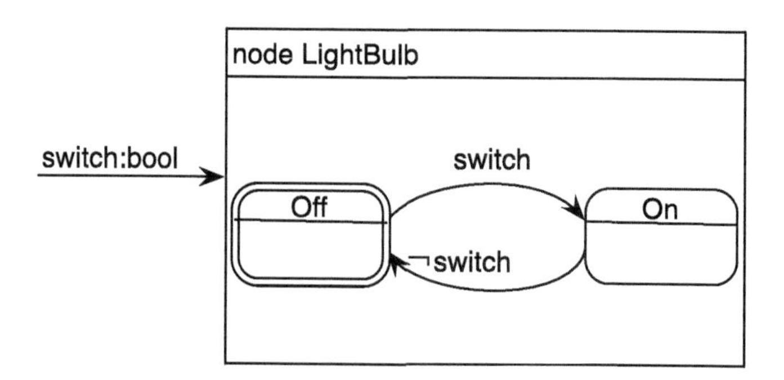

Fig. 3. Simple light bulb model.

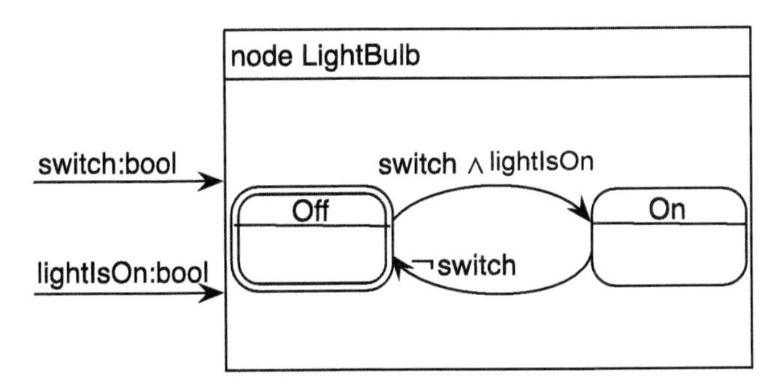

Fig. 4. Observable lightbulb.

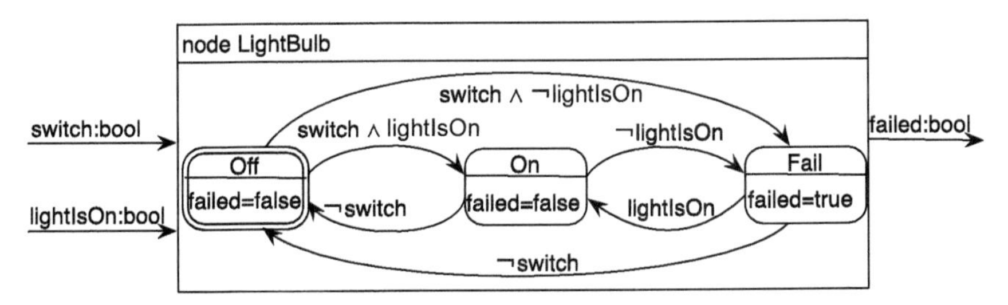

Fig. 5. Lightbulb failure model.

4.2 Person

Any person interacting with the system can be observed through its various types of sensors. However, in the very specific context of smart homes for disabled people, some characteristics of the person's behavior and impairments can also influence security rules, and thus, have to be both observable and represented in the synchronous model.

As shown in the related literature, usual observable properties about a person are position, mood, ADL and impairments. Position can be trivially defined as an LTS, depending on how rooms are interconnected in the house. Let's suppose there are three rooms: a kitchen, connected to a bathroom and a bedroom. If sensors can determine the current position of a person, then the position evolution over time can be modeled by the LTS as shown in Fig. 6. Observable behaviors in a smart home system are usually defined as a set of scenarios containing multiple steps and conditions to go from one step to another, so as to be processed by AI—in combination of events coming from the system—which can infer as to which step of which scenario the person is currently doing. Such a representation for scenarios makes them easy to be defined as LTS. And because a scenario can be aborted at any time by the person, modeling a scenario can follow the same principles presented for the observable failure of the light bulb. Figure 7 shows an LTS example representing the act of making coffee, evolving from step to step using AI notifications. Finally, mood and impairments are usually represented by boolean or numerical attributes, so that they can be represented using synchronous equations. Evaluation of impairments, for example, can come from various assessments, such as the Global Deterioration Scale for Assessment of Primary Degenerative Dementia (GDSAPD) which allocates a number between 1 and 7, depending on the cognitive decline (7 being very severe). We could also add additional disabilities such as 'blind' or 'deaf' which can be associated to booleans (Fig. 8). It should be noted that this impairment model cannot evolve as it does not take inputs to influence the person's profile. So, in the case this person is diagnosed with additional problems, this model should be changed accordingly and recompiled. But this evolution could be represented with LTS and inputs as usual. Using all these specifications, a smart home model can be completed by specific properties required by DSC, namely, designation of controllability within the model and security constraints definition.

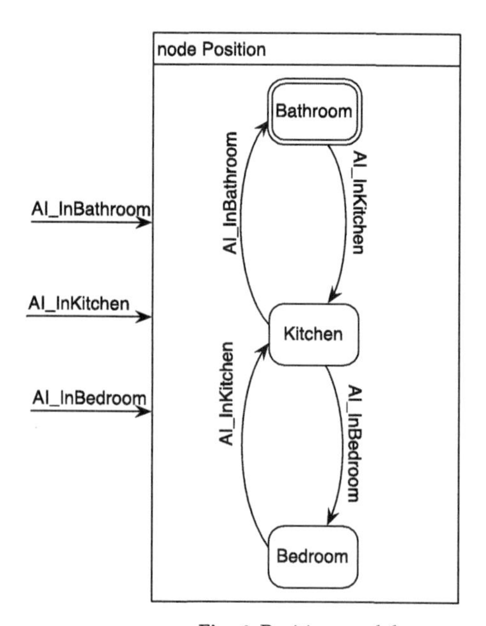

Fig. 6. Position model.

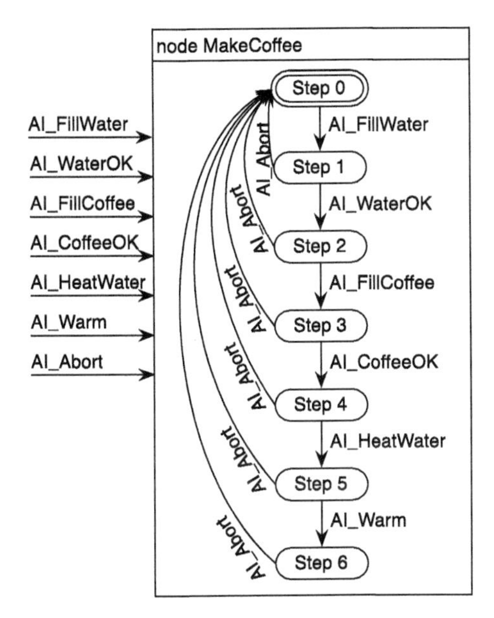

Fig. 7. ADL 'Make Coffee'.

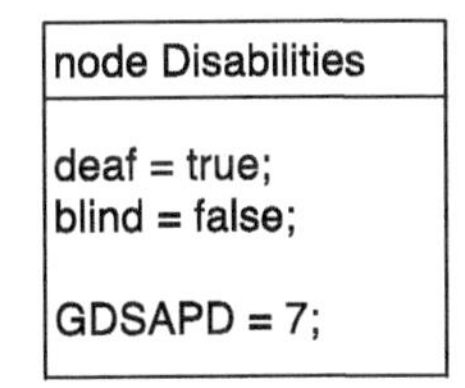

Fig. 8. Impairment model.

5. Applying DCS

When the various components and properties of a system are defined as behavior models (LTS, etc.) and synchronous equations, setting both the controllability and execution constraints enables the use of DCS.

5.1 Controllability

Controllability occurs naturally in the smart home domain. In the synchronous model, inputs are received each time the system is triggered and these can come from both the environment—uncontrollable inputs IU (e.g. a button is pressed by a human)—and the system itself—controllable inputs IC (e.g. a device is forced to shut down by control system which is part of the execution loop).

For example, let's take a system allowing a third party application to control two failure-prone light bulbs so that they can be forced to light up or remain lit even if their switch is turned off by a human. Figure 9 represents the design by constraint controller of this small system, instantiating two times the LightBulb node (modified compared to Fig. 5 with a boolean variable *c* representing the aforementioned controllability), which takes amongst others the switches values as uncontrollable inputs switch1, switch 2 ∈ IU and the values given by the third party application as controllable boolean inputs c1,c2 ∈ IC. The statement *with*, declaring controllable variables, is actually implemented in BZR, which also allows declaration of security constraints so that these variables can be valued accordingly at each instant of the synchronous execution.

5.2 Constraints

We consider two types of security constraints expressed as boolean synchronous expressions: (1) hypothesis, which is supposed to remain true for all executions, and (2) guarantee, which is enforced to remain true, using controllable variables, but only if the hypothesis stays true from the beginning of the execution. For example, let's say we want to be sure that, for all possible executions, at least one light bulb is lit up if a problem (uncontrollable information coming from observation) arises. This can be specified using the guarantee $\neg$problem $\vee$ light1 $\vee$ light2 (enforce statement). However, the system is not controllable with this rule alone: light bulbs can be in fail mode at the same time while the system receives a problem and thus the guarantee cannot be fulfilled for this specific execution. This situation would be found automatically when applying DCS, which would fail to build a controller. Now, let's say that the light bulbs can still fail but are supposed to be repaired quickly enough so that they don't fail at the same time. This is an example of fault tolerance: ultimately everything can fail but if there is enough redundancy, we can safely state that not everything will fail at the same time. The hypothesis $\neg$(fail1$\wedge$fail2) (see assume statement) represents this assumption in a synchronous boolean expression. Applying DCS, using the BZR toolset on such a model, gives back the C code of a controller taking IU as inputs and providing the computation of IC as outputs so that the system can now be executed, receiving both IU and IC. DCS is able, in this example, to find automatically

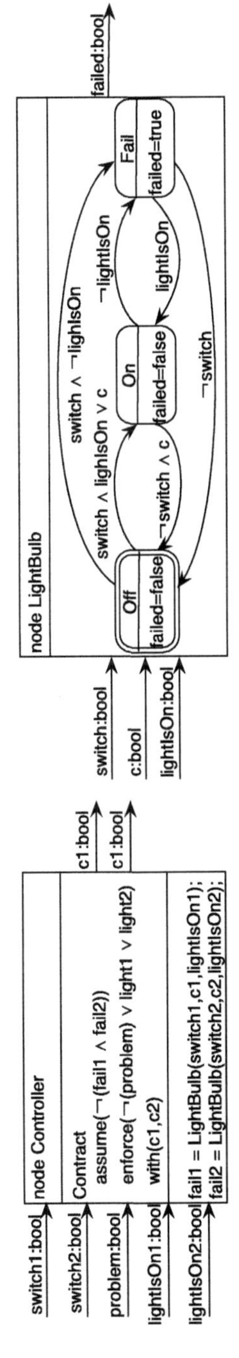

Fig. 9. Controllable light bulb model.

the correct controller code so that $c1$ and $c2$ can be valued to true or false exactly when they should (e.g. when a problem arises and lights are off, and light1 has failed, then c2 will be forced to be false, etc.). From such a minimal example, we understand how DCS becomes interesting when the system's complexity increases while having to maintain its safety. If we add other failure-prone devices, impairment models, security constraints, etc. both designing and verifying the maximally permissive controller quickly start is hard without appropriate tools.

6. Experiment

This section shows the application of DCS on the model of a smart home system to address various errors coming from the user's behavior or the system itself (failure of its components). The examples are built incrementally, i.e. they can be merged together into a model of a system on which DCS can be applied to synthesize a controller guaranteeing all user/ component safety properties. They show four types of control behavior: (1) adaptation and (2) usage limitation to anticipate a user problem (known disabilities and potential behavior errors), and (3) adaptation, and (4) usage limitation to anticipate components-related problems (hardware failure). These types of control behaviors are an answer to the smart home fault tolerance problems identified in (Chetan et al. 2005).

6.1 Base Model

Before describing the aforementioned scenarios, let's represent the base model of a smart home system (see Fig. 10). This model contains the elements concerned (directly or indirectly) by security constraints. Their behavioral definitions are given as a set of automata and synchronous equations following the BZR concepts. The model also specifies inputs and outputs, and follows the synchronous execution definition—each execution step consumes all inputs and computes all outputs through the specified equations and automata, the actual organization of computations inside a step being solved by the synchronous compiler. This specific view of the smart home system (i.e. its control-related information) constitutes the designed-by-constraint definition of the controller that we will try to synthesize. Point 1 of this model represents the main node—the controller node—centralizing all incoming events and all necessary outputs. The first four inputs (fail and repair) are related to the failure and repair events of specific elements named islands, coming from a previous implementation

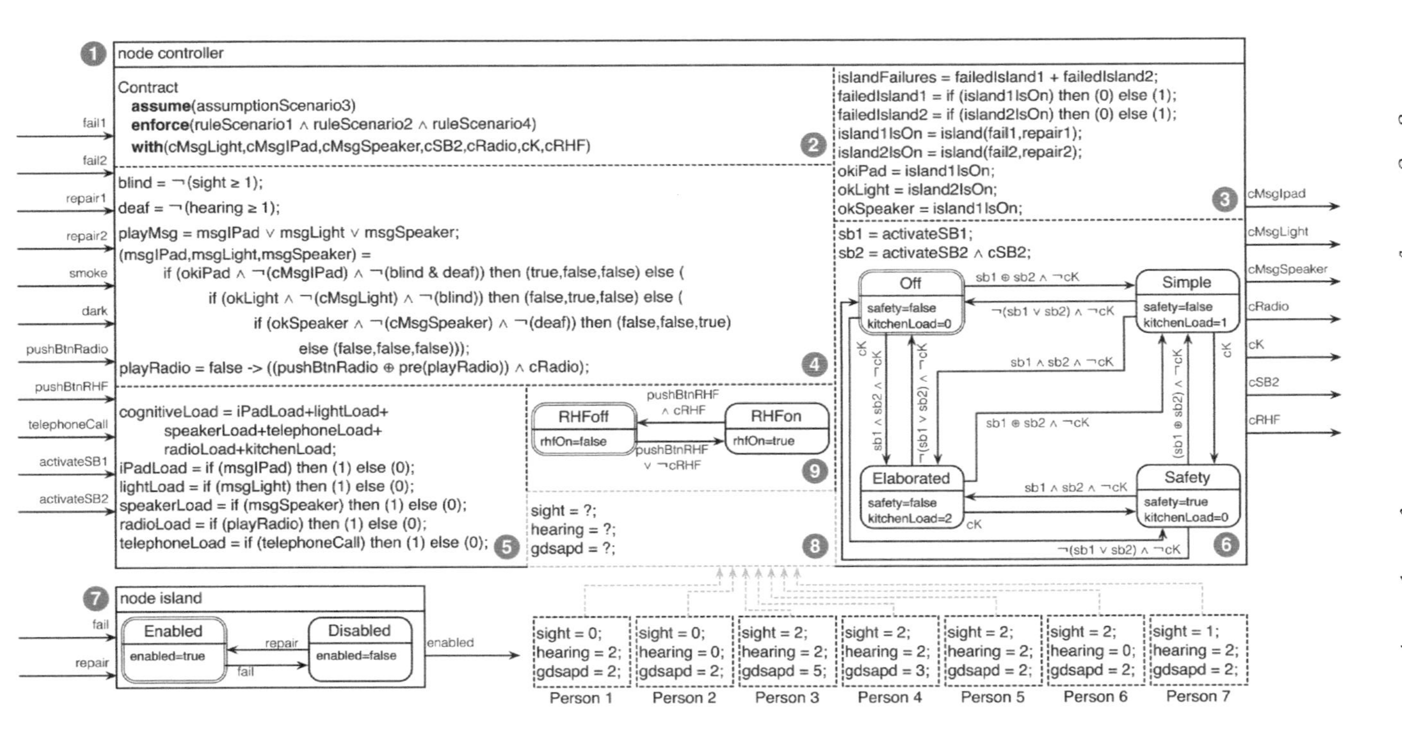

Fig. 10. Graphical representation of the defined by constraint controller.

of fault tolerance (Bouchard et al. 2012). Islands are independent systems monitoring several sensors and effectors. Here we consider two of them, instantiating (point 3) the generic node island definition given in point 7. They manage respectively (1) an iPad, a speaker, and (2) a light bulb (point 3).

Beyond island events, the controller node also receives a notification when smoke is detected (smoke), when the kitchen room is in the dark (dark), when the radio is activated (pushBtnRadio), when the range hood fan in the kitchen is activated (pushBtnRHF), when the person receives a telephone call (telephoneCall), and when the stove burners 1 and 2 of the kitchen stove are set to their ON position (activateSB1, activateSB2). Point 6 shows a simplified behavior for the kitchen stove, allowing the activation of the two aforementioned stove burners and going into a safety mode if some problem is detected. When this mode is activated, the stove burners are set to OFF and the controller is notified (through the variable named 'safety') for actions to be derived.

Point 8 represents an abstract and simplified impairment model, which will be filled with a person's actual data. Here we will see what happens with seven persons given their information about their sight and hearing (integers from 0 to 2, meaning respectively 'completely impaired' to 'normal'), and also on their GDSAPD evaluation (integer from 1 to 7, meaning respectively 'normal' to 'severely mentally disabled'). This information will influence how the system should communicate, choosing between the iPad, the speaker, or the light, and when to communicate or not—playMsg being true or false (point 4). This communication with the user will also take care of the cognitive load: communicating with the person using the iPad, etc. increments the load by one unit (point 5). If the kitchen stove is in use, it will also increase the load by one or two depending on the number of activated stove burners (kitchenLoad from point 6). Finally, receiving a telephone call and listening to the radio are also considered as cognitive loads. Thus, telephoneLoad and radioLoad (point 5) increments the load by one, depending on their activation (respectively influenced by the telephone input and the playRadio equation from point 4). Finally, automata from 9 and 10 represent the activation behavior of the range hood fan and the kitchen light.

Now the model just needs a contract (which will be detailed in the next parts), so that DCS can be applied to eventually obtain an executable controller. This contract is given in point 2 where, first, we make assumptions about the uncontrollability (assumptionScenario 3), i.e. we define synchronous equations representing some key parts of the environment's behavior. This helps DCS to eliminate the verification of event combinations and sequences that are not supposed to happen; and second, we specify rules that must remain true for all possible executions (ruleScenario 1,2,4), with the help of seven controllable variables representing the actual controllability of the system (i.e. they represent the real interface proposed by the smart home so that it can be influenced through the use of a computing system). These controllable variables are valuated internally by the synthesized controller (obtained through DCS) and provided as outputs, so that the system can react at each control step. When these variables are forced to false, cMsg(IPad/Light/Speaker) indicate which prompting system has to be used (respectively the iPad, the light and the speaker), cRadio can force the radio to turn off, cK can set the kitchen stove in a safety mode, cSB2 can prevent the use of the second stove burner and cRHF can activate the range hood fan.

We will now incrementally set four types of safety rules to tolerate errors or to adapt upon dangers and detail their effects when the smart home system under control is used by people with different disabilities and different levels of impairments.

6.2 Scenario 1: User Assistance (Adapted Prompting)

Description and objective

This first scenario aims at showing the added value of DCS when designing a controller to adapt the way the smart home communicates with the user. Guaranteeing the safety of the controller (with respect to rules) is made through the verification aspect of DCS, just like in classical formal verification algorithms (e.g. Model Checking); but knowing if such a controller actually exists, is addressed by the very specific aspect of DCS—synthesis from constraints. Let's specify a first constraint. For example, we want to be sure that when smoke is detected in the kitchen (smoke variable is true), then the range hood fan activates and a prompt information is provided through an adapted media (iPad, speaker or lightbulb). To reflect this in the model, we set the following equation for ruleScenario1:

1. ruleScenario1 = $\neg$(smoke) $\vee$ (rhfon $\wedge$ playMsg)

We now take the cases of persons 1 and 2: Will the smart home be controllable, i.e. will it be able to cope with this constraint for all possible execution? Let's apply DCS with each user profile on this specification—containing the base model, the user profile this first rule (the other ones being set to true for the moment) and let's simulate a scenario where the user activates the two stove burners to start cooking some food, but smoke is detected when the food is starting to burn.

		Person 1					
		Step 1	Step 2	Step 3	Step 4	Step 5	Step 6
Inputs	smoke	0	0	**1**	1	**0**	0
	activateSB1	**1**	1	1	1	1	1
	activateSB2	0	**1**	1	1	1	1
	pushBtnRHF	0	0	0	**1**	0	**1**
Out.	cMsgSpeaker	1	1	**0**	0	**1**	1
	cRHF	1	1	**0**	0	**1**	1

Fig. 11. Execution of scenario 1 for person 1.

Comments on scenario execution

DCS fails when applied with the profile of person 2, who is both blind and deaf, so there is no appropriate communication media in case of a problem (e.g. when smoke is detected). Technically variables msg(IPad/Speaker/Light) can never be set to 'true' whatever the values given to the controllable variables cMsg(IPad/Speaker/Light) because blind and deaf are true; thus playMsg can never be true. This leads to the fact that ruleScenario1 can be false if smoke, an uncontrollable variable given as input, is true, which is of course not permitted (ruleScenario1 has to be enforced to 'true' for all executions). Because of this, the smart home system is not controllable in this first rule and this is the reason why DCS fails, meaning that the system has to be reworked (e.g. by adding adapted medias), before person 2 can actually use it safely.

However, DCS succeeds when applied with the profile of person 1, meaning that a controller has been found so that ruleScenario1 will always remain true for all possible execution. Figure 11 shows the simulation results, highlighting important events. It is to be noted that only the relevant events and steps are represented; missing inputs are false, missing outputs are true and a new step is represented only when something changes from the previous one (new inputs/outputs or internal state modification).

- Steps 1 and 2: Person 1 activates respectively the stove burners 1 and 2 to start cooking.
- Step 3: Some smoke gets detected by a sensor that valuates the smoke input to 'true'; the controller reacts by setting the controllable variables cMsgSpeaker, cK and cRHF to 'false', which has the following effects: the range hood fan activates a message about the smoke is played using the speaker; physically, the system is able to select the appropriate message and media, knowing that smoke is true and cMsgSpeaker is forced to 'false'. This way, ruleScenario1 remains 'true'.
- Step 4: The person tries to deactivate the range hood fan. However, because smoke is still detected, the fan has to stay activated; deactivation is actually prevented by forcing cRHF to remain false in this step, avoiding to take the transition from step RHFon to RHFoff.
- Step 5: No smoke is detected anymore and no controllable variable has to be forced to 'false'. This has the following effect: the message about smoke is stopped being played.
- Step 6: The person tries to deactivate the range hood fan, which this time is permitted because cRHF is not forced to 'false'.

This scenario has shown the interest of using DCS in this context to both verify that the smart home is able to adapt to a person's disabilities and generate a controller to manage the smart home adaptation behavior.

6.3 Scenario 2: User Error Prevention (Simultaneous Devices Usage Limitation)

Description and objective

This second scenario shows the advantage of using DCS to put limitations on how the various devices in a smart home can be used, depending on the user's profile. As an example of such a case, we will focus on the compound cognitive load due to simultaneous device usage by a person and show how the smart home system gets configured to prevent a cognitive overload. Let's specify a second constraint, such that the cognitive load cannot exceed 2 and 3 units when the person's GDSAPD is respectively evaluated to 5 and 3. Regarding the adaptation possibilities, the radio can be turned off automatically and the kitchen stove cannot be used on Elaborated mode (only the stove burner 1 can be activated at most) to reduce the cognitive load when it is necessary. We set this as the following equation for ruleScenario2:

2. ruleScenario2 $=(\neg(\text{gsdapd} \geq 5) \vee (\text{cognitiveLoad} \leq 2)) \wedge (\neg(\text{gdsapd} \geq 3) \vee (\text{cognitiveLoad} \leq 3))$

Like in the example given in the first scenario, if the smart home cannot be adapted to a person (due to disabilities) for all possible executions, DCS will fail to find a controller. So we will take the cases of persons 3, 4 and 5, for which DCS succeeds and simulate a scenario where the person is listening to the radio, then activates the stove burners 1 and 2 to start cooking, but then receives a telephone call.

Comments on scenario execution

Figure 12 shows the simulation results, where we can see how the smart home gets adapted to cope with the differences in abilities to deal with cognitive load between the three persons. Person 5, having a GDSAPD lower than 3, should be able to deal with a high cognitive load, but it is not the case for persons 4 and 3, and this is why they experience some limitation when using some devices simultaneously in this controlled smart home.

- Step 1: Persons 3, 4 and 5 start listening to the radio; cognitive load is then set to 1 unit, which is correct for everyone.
- Step 2: Persons 3, 4 and 5 activate the first stove burner, increases the cognitive load by 1 unit, which is now the acceptable limit for person 3, having a GDSAPD equal to 5 units.
- Step 3: Person 4 and 5 activate the second stove burner. Now the cognitive load is set to 3 units, the acceptable limit for person 4 (having a GDSAPD equal to 3 units); but when person 2 tries to activate the second stove burner by setting the activateSB1 switch to 'on' (true), the radio goes off because the controller forces cRadio to be 'false', thus keeping the cognitive load below the acceptable limit, as required by ruleScenario2. It is interesting to note here that preventing the second stove burner to start by forcing cSB2 to be false would also have been a correct response from the controller. The order in which the controllable variables are set actually depends on the order they are declared in the BZR program. Here cSB2 is declared before cRadio, so when a value is asked for cSB2 in a step, the value of cRadio is not decided yet. In this example, cSB2 has no reason to be forced to 'false', because there is still a solution to comply with the rules (i.e. by setting cRadio to 'false'). Inverting the declarations of cSB2 and cRadio would have let the radio 'on' and prevented the use of the second kitchen stove for person 3 in this step.

		Person 3				Person 4				Person 5			
		Step 1	Step 2	Step 3	Step 4	Step 1	Step 2	Step 3	Step 4	Step 1	Step 2	Step 3	Step 4
Inputs	pushBtnRadio	**1**	0	0	0	**1**	0	0	0	**1**	0	0	0
	telephoneCall	0	0	0	**1**	0	0	0	**1**	0	0	0	**1**
	activateSB1	0	**1**	1	1	0	**1**	1	1	0	**1**	1	1
	activateSB2	0	0	**1**	1	0	0	**1**	1	0	0	**1**	1
Out.	cRadio	1	1	**0**	0	1	1	1	**0**	1	1	1	1
	cSB2	1	1	1	**0**	1	1	1	1	1	1	1	1

Fig. 12. Execution of scenario 2 for persons 3, 4 and 5.

- Step 4: A telephone call is received in this step, which increases the cognitive load by one and this cannot be prevented (there is no controllability on receiving telephone calls for this smart home, at least in the model we have defined). Person 5 can receive the call whilst continuing to cook and to listen to the radio. However, letting the cognitive load reach 4 units is not permitted for person 4, so the smart home has to react to not let this happen. This is why the radio gets deactivated (and not the stove burner 2 for the same reason as explained in step 3 for person 3). Finally, this telephone call impacts the smart home usability for person 3 to keep the cognitive load to an acceptable level. The controller forces cSB2 to be 'false' here, thus deactivating the second stove burner and keeping the cognitive load to 2 units.

This scenario shows an example of how a person's static profile can be used to prevent user errors (by keeping the cognitive load below an adapted level) by configuring the smart home with the help of DCS which provide a corresponding smart home controller.

6.4 Scenario 3: Component Failure (Redundancy)

Description and objective

This third scenario shows the application of DCS to solve a problem that could arise from a previous implementation of fault tolerance in the smart home system: in our architecture presented in (Bouchard et al. 2012), we "did install industrial grade material... to avoid hazardous situations, for example, where the resident cannot turn on the light due to a system failure." We connected our various sensors and effectors to four independent fault-tolerant islands so that "if a block falls, only the [connected equipments] will be affected." Sensors and effectors are critical safety elements, so if their connected island fails, do we have enough redundancy? To generalize, having enough redundancy in a given system means that for all possible execution of this system and in case of a failure, there is always a solution to keep it running correctly. It means here that the smart home system remains adapted to the person's impairments. In order to keep the base model small and visually clear, we consider simplification of our own redundancy implementation where only islands can fail but not the other devices (lights, sensors, etc.). Island failure is a kind of uncontrollable event (the system can do nothing to prevent this), so it cannot be represented as a control rule. However, we assuming a case where two islands are disabled in the same step should never happen, we will tolerate one and only one failure at most. This hypothesis can be represented in the assume part of the contract, by giving the following equation to assumption Rule3:

3. Assumption Rule3 = Island failures ≤ 1

In this scenario, we will see that having an island failure may have different impacts on the smart home behavior depending on who is actually living inside. It starts when the island number 1 fails. Then the user activates the two stove burners to start cooking, but smoke gets detected. This triggers a message (using an appropriate media) and the user deactivate the stove burners. At some point, no more smoke is detected and the island number 1 gets repaired. Then the user re-starts cooking, but smoke gets detected again and a new message has to be communicated. We take persons 1 and 6 for this scenario and apply DCS.

		Person 6							
		Step 1	Step 2	Step 3	Step 4	Step 5	Step 6	Step 7	Step 8
Inputs	fail1	**1**	0	0	0	0	0	0	0
	repair1	0	0	0	0	0	0	**1**	0
	smoke	0	0	0	**1**	1	**0**	0	**1**
	activateSB1	0	**1**	1	1	**0**	0	**1**	1
	activateSB2	0	0	**1**	1	**0**	0	**1**	1
Outputs	cMsgLight	1	1	1	**0**	0	**1**	1	1
	cMsgIPad	1	1	1	1	1	1	1	**0**
	cMsgSpeaker	1	1	1	1	1	1	1	1

Fig. 13. Execution of scenario 3 for person 6.

Comments on Scenario Execution

For person 1, DCS does not find an appropriate controller. Indeed, if island number 1 fails, then the speakers cannot be activated (Point 3, island1IsOn being false means okSpeaker is false too and then msgSpeaker cannot be true). However this is the only acceptable communication media when a person is blind (but not deaf) which is the case of person 1 (Point 4); if blind is true, then msgLight and msgIPad cannot be true; so if msg(IPad/Speaker/Light) are all false, then playMsg becomes false and this can be problematic: if a message has to be communicated to the user because of a problem, such as a smoke problem as specified in ruleScenario1, there is no media available and because this situation cannot be prevented by any available controllability then DCS fails, meaning that the redundancy implementation has to be reworked.

For person 6 however, DCS succeeds and the results of the aforementioned scenario can be seen in Fig. 4. It has to be noted that if both the islands fail at the same time, then no media can be selected any more, and in case of a smoke problem, this would violate ruleScenario1. But DCS ignores this case, as we have defined that a double island failure should not happen at the same time in the assume part of the contract.

- Step 1: The scenario starts when island number 1 fails; this disables the use of the iPad and the speaker as communication devices.
- Steps 2 and 3: Person 6 activates the two stove burners to start cooking.
- Step 4: Smoke gets detected from the environment (this enables the range hood fan but it is not relevant in this scenario), and a message has to be communicated in order to keep ruleScenario1 to true. Person 6 being deaf, the iPad cannot be used as it requires to have correct sight and hearing (Point 4); so only the light remains, and this is why cMsgLight is forced to false by the controller. This way, a message about the smoke problem can be communicated through the light.
- Step 5: Person 6 sets the two stove burners' switches to "off".
- Step 6: No smoke is detected, so the controller does not continue to keep cMsgLight to false (having smoke to false keeps ruleScenario1 to true). Thus the message about smoke can be stopped.
- Step 7: Island 1 has been repaired and the two stove burners are switched "on" again.
- Step 8: Smoke is detected anew: this time, the message about smoke gets communicated through the iPad, because (1) island number 1 is operational and (2) the controllable variable cMsgIpad is declared after cMsgLight, explaining why cMsgLight remains true (not forced by the controller), because cMsgIpad can always be forced to false after (which is the case in this step), thus keeping ruleScenario1 to true.

Use of DSC in this scenario is especially powerful: instead of trying to define redundancy generically for all types of users, we can reduce costs by defining more or less redundant component combinations for different users and applying DCS to guarantee that a specific redundant installation is safe for a specific user profile. Moreover, adjusting hypothesis on the components' quality is simply a question of setting a boolean equation in the assume part. For example, if we have five islands monitoring our components, and want to tolerate a maximum of three failures at the same time and see if our system is still stable, we just have to set (islandfailures ≤ 3) in the assume part and apply DCS to know if the smart home is actually stable (controllable) in this context and get the associated controller.

6.5 Scenario 4: Component Adaptation (Degraded Mode)

Description and objective

An alternative solution to cope with hardware failures, besides using redundancy, is to modify the way the remaining operational components are used. If we take the example of islands failures, and want to cope with a double failure, which creates a communication problem when smoke is detected, then we can program a controller to degrade what can create smoke (i.e. the kitchen stove by setting cK to false), and either remove ruleScenario1 or assume that smoke remains false when Safety mode is active. But it would mean that the kitchen stove would be forced to remain in Safety mode for all possible executions just because smoke could happen, which is not acceptable. Instead, we want to find an example showing how DCS can be useful to build a controller helping to anticipate a hardware failure by modifying the remaining active components in the case where no redundancy is available. Let's say we only have one lightbulb in the kitchen. At night, if the lightbulb fails—and whatever its controllability (i.e. the lightbulb can be switched "on" or "off" automatically)—then the kitchen goes completely dark. Now let's build an new safety rule such that if the user has limited sight, the kitchen stove cannot be used at night when the lightbulb is not activated. Of course, because the light bulb has no redundancy, we cannot create a rule such that when the kitchen is used at night, then the light should go "on" if sight is limited. This model would not comply with the reality if we consider that the lightbulb can fail (and thus cannot go "on" at some undetermined moment). So instead, we use a light sensor providing the value of a variable named dark—indicating if there is enough light in the room (false) or not (true)—and this value explicitly impacts the way the kitchen can be used through the following synchronous equation attributed to ruleScenario4:

4. ruleScenario4 = $\neg(\text{sight} == 1) \vee (\neg\text{dark}) \vee (\text{kitchenLoad} == 0)$

This means that if the user's sight is evaluated to 1 (visually impaired but not blind) and if the kitchen is in the dark, then the smart home has to adapt itself so that the kitchen stove cannot be used (it is either in Off or Safety mode, the only modes where kitchenLoad equals 0).

		Person 7				Person 5			
		Step 1	Step 2	Step 3	Step 4	Step 1	Step 2	Step 3	Step 4
In.	dark	1	1	0	1	1	1	0	1
	activateSB1	0	1	1	1	0	1	1	1
O.	cK	1	0	1	0	1	1	1	1

Fig. 14. Execution of scenario 4 for persons 7 and 5.

We now define a scenario where we start at night, the kitchen is in the dark and the user activates the stove burner 1. Then the user presses the light button (this event is not given to the controller because no information about the actual lightbulb activation can be safely derived from it), but the lightbulb fails in the following instant.

Depending on the user's profile, different control results happen when this scenario is played, as shown in Fig. 5 for persons 7 and 5 (for which DCS succeeded). This example relates to Scenario 1 regarding the explicit constraint definition on environment events (smoke value directly impacts the activation of the range hood fan), to Scenario 2 regarding the dynamic adaptation to multiple users profiles and to Scenario 3 regarding the hardware failure example (islands can fail, their failures impact the system's behavior).

Comments on scenario execution Person 7 being visually impaired, the smart home system adapts itself consequently. However, person 5 with good sight, the smart home does not interfere with the kitchen stove use, whatever the light condition.

- Step 1: Both users are in the kitchen, in the dark.
- Step 2: Both of them press a button to activate the stove burner 1, but the actual activation is prevented for person 7. The controller forces cK to false, which prevents the kitchen stove to go in Simple execution mode where kitchenLoad would be equal to 1 instead of 0, thus violating ruleScenario4.
- Step 3: They both activate the lightbulb, and the light sensor reacts by setting dark to false. This allows the actual activation of the stove burner 1 for person 7.
- Step 4: The kitchen returns to dark, as indicated by the dark value (true), even if neither person 7 nor 5 touched the light switch to turn it off. The lightbulb has failed and the controller reacts on the dark

value—instead of the actual light switch position—for person 7 by setting the cK controllable variable to false, thus placing the kitchen stove to its safety execution mode.

In the context of usage limitation to anticipate components-related problems, this example shows again the advantage of being able to define the system's controllability by constraint, instead of giving its actual implementation. Here the kitchen stove is actually controllable in the sense that an internal system (the controller) can act on it to prevent the activation of its stove burners, but the kitchen model does not have to define explicitly under which conditions it has to react; the actual control implementation (valuation of the controllable variables) is obtained by synthesis, given the global constraints defined by the programmer in the model of the system's components. States or behaviors of one or several components (e.g. the light and impairment profile) can have an impact on the controllability of other components (e.g. the kitchen stove) without requiring to define explicitly this controllability.

6.6 Evaluation

Earlier we had implemented some redundancy mechanisms in our own smart home test lab (Bouchard et al. 2012), so that not all sensors/effectors would be controlled by a single machine (an Island) because it would have consisted in a single point of failure. However, we could not be exactly sure that for every susceptible failure, the use of redundant elements (islands) would be sufficient to keep the smart home safe for a particular person (as several sensors and/or effectors not managed by the new island would have been disabled). The way we connect our sensors and effectors to multiple islands could actually be safe for a person in case of a failure, but not for another one with different disabilities. This could be hard to find and verify without appropriate tools to solve this combinatorial problem.

Redundancy without verification is indeed not sufficient. While trying to solve this failure problem, we compared our approach to closely related ones regarding smart homes for disabled people and learned, especially from (Chetan et al. 2005), that failures could be of multiple types in this context and redundancy itself was not the only solution to address them. This is why we became interested in the more general problem of fault tolerance for such smart homes and we would base our use cases on their studies to show how different types of failures could be addressed. But unlike these approaches, we would complete ours by verifying it.

Designing smart home models such that they can be verified has been done several times. One of the most related approaches regarding modeling and verification is (Corno and Sanaullah 2013), where a smart home is modeled using a formal representation (State Charts) and safety properties defined so that formal verification tools can be employed to guarantee that these properties are ensured for all possible execution. However, we already discussed the problem of modeling the entire system to apply verification. That's why we kept aside the formal modeling approach, but took an approach based on synthesis to address this combinatorial problem by solving it automatically from constraints, instead of trying to find (and verify) a complete solution manually. Our expertise with synthesis techniques came from previous work in the domain of reconfigurable hardware architecture, where DCS was proven useful to build formal reconfiguration controllers. Still, in the hardware context, DCS was also employed to carry out computations on failure-prone processors, giving us inspiration on how to actually use DCS to manage fault-tolerant smart homes.

Usage of synthesis techniques in the smart home context is very recent. First results can be seen in (Zhao et al. 2013) and (Guillet et al. 2013) which present the use of DCS respectively from a generic point of view (in the context of the Internet of Things) and from a specific one regarding fault tolerance. However, being preliminary, they both lack concrete use cases, implementation and results. This proposal makes a contribution over them by presenting these elements: a detailed methodology to create smart home controllers by synthesis, using BZR for smart home elements modeling and use cases with realistic scenarios that were tested in our lab to demonstrate the relevance and advantages of using DCS for addressing fault-tolerance problems (identified in (Chetan et al. 2005)) that may occur in a smart home dedicated to disabled people.

7. Conclusion and Perspectives

Safety and security services are essential requirements for many pervasive computing systems. This is especially true for smart homes dedicated to people with disabilities, where security constraints prevail. They represent a pervasive system category where safety is actually a very critical property: the person living in such a house is usually frail and is not supposed to be able to cope with errors. Implications of failures can range from user annoyance to hazardous situations.

Correct adaptation behavior so that the smart home remains safe, whatever the conditions of execution, is both difficult to design and verify. While verification has been addressed multiple times, use of synthesis techniques in this context to guarantee a safe behavior (employing formal verification) while simplifying the design (which is derived from constraints) is still rarely encountered and lack of examples show how they can be used to solve practical problems. In this context, this proposal makes a contribution by providing a design methodology, relying on DCS and backed by scenarios examples, to build smart home controller systems guaranteeing safety properties.

The results validating the proposal present both modeling and executions parts for different scenarios. They especially focus on fault tolerance as a safety property and show how to deal with four types of typical control needs in this context: adaptation and usage limitation for users' problems and components failures. With these results, obtained by rigorous experiments (real scenarios executed in our smart home test lab), we demonstrated that the synchronous paradigm (on which BZR is based) and DCS tools (such as Sigali) are relevant to design and compute the controller of a smart home system, in the context of fault tolerance.

In the end, the proposed methodology allows us to solve a simple but crucial question—Can this smart home be adapted to this person, for every failure situation that can be derived from its model? A negative answer implies that a safety constraint (defined in the model) can be violated and this cannot be prevented. The smart home itself has to be modified (by adding more redundancy, removing dangerous elements or executions modes, etc.) because no correct adaptation controller exist. However, a positive answer to this question automatically gives back the code of a correct control system to be connected (inputs and outputs) and executed within the corresponding smart home so that it can actually be adapted dynamically.

As a perspective, the current methodology could be improved by defining an adequate abstraction level so that smart home designers would not even have to learn about BZR. For example, such an abstraction has been implemented in the reconfigurable embedded systems domain (Bouhadiba et al. 2011) to allow designers to specify reliable reconfiguration controllers, using only a UML profile (high-level abstraction). Models built with this profile could be transformed into a synchronous representation based on BZR to make DCS applicable transparently, thus giving back the executable code of their specified-by-constraints controllers.

Keywords: Smart Homes, Cognitive deficit, Controllability, Discrete Controller Synthesis, Fault-tolerance

References

Augusto, J.C. and C.D. Nugent. 2006. Smart homes can be smarter. *In*: In Designing Smart Homes—The Role of Artificial Intelligence.

Benveniste, A., P. Caspi, S.A. Edwards, N. Halbwachs, P. LeGuernic and R. deSimone. 2003. The synchronous languages 12 years later. Proceedings of the IEEE. 91: 64–83.

Bouchard, B., S. Giroux and A. Bouzouane. 2007. A keyhole plan recognition model for Alzheimer's patients: first results. Journal of Applied Artificial Intelligence (AAI). 21: 623–658.

Bouchard, K., B. Bouchard and A. Bouzouane. 2012. Guidelines to efficient smart home design for rapid AI prototyping: a case study. *In*: Proceedings of the 5th International Conference on Pervasive Technologies Related to Assistive Environments, New York, NY, USA, ACM.

Bouchard, K., B. Bouchard and A. Bouzouane. 2013. Discovery of topological relations for spatial activity recognition. pp. 1–8. *In*: Proceedings of the IEEE Symposium Series on Computational Intelligence (SSCI 2013).

Bouhadiba, T., Q. Sabah, G. Delaval and E. Rutten. 2011. Synchronous Control of Reconfiguration in Fractal Component-based Systems—A Case Study. Proceedings of the International Conference on Embedded Software. EMSOFT.

Bulow, J. 1986. An economic theory of planned obsolescence. The Quarterly Journal of Economics. 101: 729–749.

Carberry, S. 2001. Techniques for Plan Recognition. User Modeling and User-Adapted Interaction. 11: 31–48.

Cassandras, C.G. and S. Lafortune. 2006. Introduction to Discrete Event Systems. Springer-Verlag.

Chetan, S., A. Ranganathan and R. Campbell. 2005. Towards fault tolerance pervasive computing. Technology and Society Magazine. 24.

Corno, F. and M. Sanaullah. 2011. Formal verification of device state chart models. *In*: 7th Int. Conf. Intelligent Environments.

Corno, F. and M. Sanaullah. 2013. Modeling and formal verification of smart environments. Security and Communication Networks.

Delaval, G. and E. Rutten. 2010. Reactive model-based control of reconfiguration in the fractal component-based model. CBSE.

Delaval, G., H. Marchand and E. Rutten. 2010. Contracts for modular discrete controller synthesis. pp. 57–66. *In*: Proceedings of the ACM SIGPLAN/SIGBED 2010 Conference on Languages, Compilers, and Tools for Embedded Systems, New York, NY, USA, ACM.

Dumitrescu, E., A. Girault, H. Marchand and É. Rutten. 2010. Multicriteria optimal discrete controller synthesis for fault-tolerant real-time tasks. pp. 366–373. *In*: Workshop on Discrete Event Systems, WODES'10, Berlin, Germany.

Fortin-Simard, D., K. Bouchard, S. Gaboury, B. Bouchard and A. Bouzouane. 2012. Accurate passive RFID localization system for smart homes. pp. 1–8. *In*: IEEE 3rd International Conference on Networked Embedded Systems for Every Application (NESEA).

Guillet, S., B. Bouchard and A. Bouzouane. 2013. Correct by construction security approach to design fault tolerant smart homes for disabled people. *In*: EUSPN.

Guillet, S., F. de Lamotte, N. Le Griguer, É. Rutten, G. Gogniat and J.P. Diguet. 2012. Designing formal reconfiguration control using UML/MARTE. pp. 1–8. *In*: 7th International Workshop on Reconfigurable Communication-centric Systems-on-Chip (ReCoSoC).

Jaygarl, H., A. Denner and N. Pham. 2008. Software Requirements and Specication Document for Smart Home Notication and Calendering System. Research Report (Smart Home Project, Iowa State University). 1–45.
Kephart, J.O. and D.M. Chess. 2003. The vision of autonomic computing. IEEE Computer.
Kilgore, C., M. Peitz and K. Schmid. 2004. System Requirements Document for Safe Home. Research Report, Iowa State University.
Lapointe, J., B. Bouchard, J. Bouchard, A. Potvin and A. Bouzouane. 2012. Smart homes for people with Alzheimer's disease: adapting prompting strategies to the patient's cognitive profile. *In*: Proceedings of the 5th International Conference on Pervasive Technologies Related to Assistive Environments, New York, NY, USA, ACM. 30: June 06–08, 2012.
Latfi, F., B. Lefebvre and C. Descheneaux. 2007. Ontology-based management of the telehealth smart home, dedicated to elderly in loss of cognitive autonomy. pp. 1–10. *In*: Workshop on OWL: Experiences and Directions.
Le Lann, G. 1996. The Ariane 5 Flight 501 Failure—A Case Study in System Engineering for Computing Systems. Technical Report, REFLECS—INRIA Rocquencourt.
Marchand, H. and M. Samaan. 2000. Incremental Design of a Power Transformer Station Controller using a Controller Synthesis Methodology. IEEE Trans. Software Engin. 26(8): 729–741.
Marchand, H., P. Bournai, M.L. Borgne and P.L. Guernic. 2000. Synthesis of discrete-event controllers based on the signal environment. Discrete Event Dynamic Systems. 10: 325–346.
Mihailidis, A., J. Boger, M. Canido and J. Hoey. 2007. The use of an intelligent prompting system for people with dementia. Interactions. 14.
Novak, M., M. Binas and F. Jakab. 2012. Unobtrusive anomaly detection in presence of elderly in a smart-home environment. *In*: ELEKTRO.
Patterson, D.J., H.A. Kautz, D. Fox and L. Liao. 2006. Pervasive computing in the home and community. pp. 79–103. *In*: Bardram, J.E., A. Mihailidis and D. Wan (Eds.). Pervasive Computing in Healthcare. CRC Press.
Picard, R.W. 1995. Affective Computing. Technical report.
Pigot, H., A. Mayers and S. Giroux. 2003a. The intelligent habitat and everyday life activity support. *In*: 5th International Conference on Simulations in Biomedicine, Slovénie, 5th International Conference on Simulations in Biomedicine.
Pigot, H., B. Lefebvre, J.G. Meunier, B. Kerhervé, A. Mayers and S. Giroux. 2003b. The role of intelligent habitats in upholding elders in residence. pp. 497–506. *In*: 5th International Conference on Simulations in Biomedicine.
Ramadge, P.J.G. and W.M. Wonham. 1989. The control of discrete event systems. Proceedings of the IEEE. 77: 81–98.
Ramos, C., J.C. Augusto and D. Shapiro. 2008. Ambient Intelligence: The Next Step for Artificial Intelligence. Intelligent Systems, IEEE. 23.
Schmidtke, H.R. and W. Woo. 2009. Towards ontology-based formal verification methods for context aware systems. pp. 309–326. *In*: Proceedings of the 7th International Conference on Pervasive Computing, Berlin, Heidelberg, Springer-Verlag.
Zhao, M., G. Privat, É. Rutten and H. Alla. 2013. Discrete control for the internet of things and smart environments. *In*: 8th International Workshop on Feedback Computing, in Conjunction with ICAC 2013.

5

Context-Aware Service Provision in Ambient Intelligence

A Case Study with the Tyche Project

Charles Gouin-Vallerand

1. Introduction

Smart environments are, since the past twenty years, an important topic of research in computer science. The major reason is that smart environments offer solutions to several problems that the modern society faces. For instance, with the rising proportion of the elder's population in most of the Occidental and Asian countries, the scarcity of caregiving resources call for a new vision in caregiving. Ubiquitous technologies, such as smart homes, give the technological support to ensure personal care at home (Rialle et al. 2008). On the other hand, the smart cities initiative corresponds to the approach, where the urban spaces are instrumentalized with sensors, citizen are provided data through crowdsensing and the collected data are analyzed to increase the effectiveness of the management and propose

Télé-Université du Québec, 5800 St-Denis Boul. Office 1105, Montreal, Quebec, Canada, H2S 3L5. Email: charles.gouin-vallerand@teluq.ca enhanced services to the citizen (Ratti and Townsend 2011). Both kinds of smart environments have their own requirements and purpose; however, it is essential to support users across these environments, notably for people with special needs.

In a way, to assist the users in their daily living activities, context-aware and intelligent systems are required to provide assistive services to users on their devices—smart phones, tablets, desktop computers or embedded devices, depending on the available devices and environment type. By context-awareness, we mean the ability of a system to capture, model and use specific information about the environment surrounding the system, such as location, time, user profile (Ryan et al. 1997). For instance, a context aware system can host software components that infer synthetic context from the raw context provided by sensors and from other synthetic contexts (e.g. other devices). Context awareness enables such a system by assisting users in performing daily-life activities or warns specialized personnel that human intervention is required. Software components can consume context, produce context for others to consume, or use context to decide upon an application domain-dependent course of action.

Numerous efforts have been made in the development of platforms to support context-awareness for ambient intelligence (Dey et al. 2001; Preuveneers et al. 2004). Most applications and studies today rely on smart spaces, i.e. physical locations equipped with a set of sensors and actuators where the basic physical layout is known beforehand. These spaces include any controlled environment where context-awareness plays a role, such as assisting people with disabilities (e.g. hospitals, hotel rooms, apartments, houses, classrooms). Thus, context-aware services can provide several benefits to people with special needs (PwSN) and a number of projects proposed in the last years present solutions that increase the quality of life of PwSN. For instance, Giroux et al. (2008) proposes a framework to support people with cognitive deficiencies, by monitoring the current state of users' activities through context awareness and assisting users step-by-step in their activities when errors or confusion are detected. Skubic et al. (2012) uses contextual information from smart home sensors to continuously monitor users' activities and assessing health changes, such as cognitive decline. Moreover, context awareness is often implemented in user mobility scenarios by using mobile devices, such as smart phones, embedded sensors and location acquisition system, e.g. GPS. For instance, Hoey et al. (2012) uses contextual information from smart phones to recognize wandering behaviors of people with dementia. A large number of other projects and publications propose solutions for PwSN based on context awareness and the three last examples give an overview of the possibility of context awareness in assisting and helping PwSN.

An intelligent service provision system allows dynamic, fast and adapted service deployment toward the users in the environments, based on the context of the environment and takes into account different constraints, such as the users' capabilities and preferences. The main goal of the proposed service provision system is to support the deployment of assistive services in the smart environments for other smart systems, like activity recognition or errors and failures recognition systems (Roy et al. 2007). These systems use the service provision functionalities by sending a deployment request to the service provision system by supplying the needed information related to the assistance that needs to be deployed: Which user? Which software? What are the software needs? Is there a specific zone of the environment that is targeted by the assistance request? There are several benefits from encapsulating the service provision into a different system than the recognition software. By using a Service Oriented Architecture (SOA), the service provision functionalities of the system can be used by several systems in a smart environment. Thus, the complexity of the provision reasoning processes are hidden for other environment's software (like in the facade design pattern) and it is even possible to have several service provision systems (or services, thanks to the SOA mechanisms) for different kinds of provision needs.

To do so, a directive or recommend-based service provision approaches are available, depending on several factors: context, type of services to deploy, user profiles, type of devices, etc. However, the complexity of the smart environments with their heterogeneous devices, specific configurations and the important quantity of information to process convert the service provision into a serious challenge when dynamism and context precision are some of the system's requirements. Thus, building and deploying context-aware service provision systems in smart environments, such as smart homes or smart urban environments is not something easy, for instance, its implementation and management of the implicit complexity, caused by the important number of components (e.g. software or devices) and their heterogeneity. The complexity of the smart spaces is similar to the problems of large enterprises that own several servers and large applications on them (Talwar et al. 2005). Deploying systems that will provide a 'plug and play' way for service provision in smart environments, by managing all the configuration and device heterogeneities, will help in a broader deployment and usage of the smart environment technologies.

This chapter will focus on the topic of the service provision, more specifically software services (e.g. assistive software) in smart environments, such as smart homes and smart urban environments. Therefore, the first section presents a review of the literature from the first work on topic to the current state of the art. This review will focus on three groups of work—the researches around the service provision and interaction delivery, the self-organization and configuration of smart environments and the context-aware recommended systems, which in a way provide services to users and devices. The second section presents some technologies that can help in supporting the service provision—the OSGi framework, the OCAP platform and the Android operating system. The third section presents a case study of a middleware for providing dynamically services in smart environment, while coping with the complexity and heterogeneity of the smart spaces: the Tyche middleware. We conclude this chapter with a discussion on the emerging trends.

2. Review of Literature

In the first definition of ubiquitous computing by Mark Weiser (Weiser 1993), provision of services to the user was implicit with its 'hundreds of computer per room'. By getting several tabs, pads and other interaction objects in an environment, the existence of algorithms that cope with the analysis of the current context and manage the deployment/running of the user interface was clearly a key element.

However, the first works to describe the service delivery or provision in smart environments were published around the beginning of this century, such as the Microsoft's EasyLiving project (Shafer et al. 1998), Carnegie Mellon's Aura project (Sousa and Garlan 2001) or the BASE project (Becker et al. 2004).

3. Service Provision and Interaction Delivery in Smart Environments

The EasyLiving project (Shafer et al. 1998) is a well-known project from Microsoft Research about the development of technologies dedicated to smart spaces. About the service provision, the EasyLiving Geometric Model (EZLGM) proposes a mechanism that determines which devices, in a given environment, can be used by a user during human-machine interactions and help in the selection of the right devices. The EZLGM models the relation (with *measurements*, which describe the position and the orientation of an entity's coordinate frame) between *entities* and the EZLGM can also represent entities' expense with the *extent* concept. Them, the EZLGM uses geometrical transformation to determine if there is a relationship between entities. If the EZLGM can manage the spatial context of a smart environment, it doesn't take into account a more complex environment context with user capabilities, preferences, device resources and capabilities.

The Aura project (Sousa and Garlan 2001) from Carnegie Mellon University was research on the transparent delivery of human-machine interaction in ubiquitous computing environments. The authors propose a framework that supports the interaction with users while they are in mobility in an environment. Their work is based on the concept of a personal Aura, which describe the user profile, its current activity, etc. and support the deployment of several types of interaction on different types of devices (e.g. Unix and Windows systems). Therefore, the Aura framework includes five modules that support the interaction provision to the users : "[...] first, the Task Manager, called Prism, embodies the concept of personal Aura. Second, the Context Observer provides information on the physical context and reports relevant events in the physical context back to Prism and the Environment Manager. Third, the Environment Manager embodies the gateway to the environment; and fourth, Suppliers provide the abstract services that tasks are composed of: text editing, video playing, etc. From a logical standpoint, an environment has one instance of each of the types: Environment Manager, Context Observer and Task Manager" (Sousa and Garlan 2001).

In (Becker et al. 2004), Becker et al. present PCOM, a component-based framework that was developed for the BASE project, which allows:

- The deployment of small applications on resource-constrained devices with the J2ME virtual machine
- The management of the app's life cycle
- The dynamic adaptation of the device's software, depending on the environment's available services

The originality of PCOM is based on the adaptation mechanism of the system and the utilisation of 'contracts' to describe the devices' needs in system components. Therefore, each component (software module that regroups services and computation processes) defines, with a 'contract', the services that are exported and the services that are required by the component to work. For instance, a component could require an input device, such as a keyboard. This input device could be replaced by an

SMS service to provide text inputs, if no keyboard is available. If the required components are not available, PCOM can hold the component until a component, giving the required service, is available. The adaptation strategies of PCOM, built by the evaluation of the 'contracts', can be pre-defined by developers for specific situations (e.g. a system which requires a specific keyboard model). The work of Becker et al. was among the first to propose a flexible and self-configurable system to support the provision of services among a collection of ubiquitous computing devices.

The O2S system (Paluska et al. 2003) proposes a solution similar to PCOM. The innovation of the system lies in the utilization of a decision tree to describe the states that must be met to trigger the deployment of software components. In this decision tree, it is possible to describe the software dependencies towards external components, join actions to the states, such as deploying specific components, verify and validate some contextual informations, etc. For each state defined in the tree, an action plan ('Planlet') is linked. During the state evaluation, O2S verify the plan viability if the constraints are respected with a reasoning engine developed in PROLOG. Unlike PCOM, O2S allow to integrate, more easily, constraints based on the contextual information (e.g. a user in a location triggers a service).

The European project *AMIGO* proposes a framework for a smart environment that enhances the assistance of users through context-awareness. In the context of this project Vallee et al. (2005) proposes a system to dynamically create end-user services through services composition. The service composition is initiated by abstracted plans which describe the environment state/context the plans are responding to, which actions to take and the notification to make to the users. These different plan steps are matched with services in the environment through a composition manager. At some points, the user profile is considered by taking into account the possible handicaps of the users.

4. Self-configuration and Organization of Smart Environments

Works on the self-configuration of software in smart environments include several aspects of service provision. As part of the Autonomic Middleware for Ubiquitous Environments (AMUN) project (Trumler et al. 2004), the authors propose a middleware to facilitate the management and deployment of software components in smart environments by integrating autonomic computing-based features. AMUN integrates a control loop more or less based on MAPE-K approach of the autonomic computing (Kephart and Chess 2003) with:

- A system event-monitoring module
- A knowledge base divided into three categories: information specific to applications (deployed applications, resources), information specific to the events from the monitored items (past events) and metric on the same system (CPU usage, network usage, etc.)
- Some control algorithms
- A configuration module using the system information and control algorithms to implement the measures in response to the sensed events

The autonomous processes between AMUN devices are based on a strategy of choreography, where each entity has its own environmental manager of autonomy (control loop), which includes grafted modules for managing communications with partners. The application deployment process of the participating nodes to the middleware is managed by the set of nodes, using a negotiation protocol. AMUN was used, among others projects, to deploy software in a Smart Doorplate project—a project offering information and assistance on screens installed on corporate office doors.

Trumler et al. (2006) describes in detail an original negotiation protocol based on social behaviors in the distribution of tasks within a group. Under this protocol, a coordinating entity (itself being one of the participating nodes) distributes the list of applications that need to be deployed unto the environment's entities. They then evaluate their ability to run each application and, in turn, communicate its capabilities to other entities via the entity coordinator. Then, the coordinator node dispatches the software components that need to be deployed in the environment based on the entity's availability.

As part of the Gaia project, Anand Ranganathan and Roy Campbell of the University of Illinois at Urbana-Champaign also offer computer-based mechanisms that deploy automatically software components in a smart environment (Ranganathan and Campbell 2004). Given the complexity of an environmental software configuration, they offer a solution based on the STRIPS planning algorithm. This algorithm is used to find a deployment solution based on user needs, before starting multimedia on devices (displays, speakers, audio players, etc.) present in a conference/presentation room. Based on a strategy of orchestration, an entity manages the configuration; adapts the method proposed in GAIA, with several changes in the internal mechanics, to other contents as media, such as applications, services and modules.

Ranganathan et al. (2005) also proposes a second strategy for automatically managing the software provision in a smart environment and which is based on ontologies and semantic matching. In this solution, the goals of users are first adapted to the context of space. For example, if the user expresses the need to control a media from a mobile device, the system reflects this need through a list of mobile devices capable of meeting the need. Once the list is produced, a semantic comparison is made between the specific needs of the user and descriptions of entities from the list. From this comparison emerges the configuration that must take the intelligent space to meet the needs of the users.

Syed et al. (2010) propose an architecture for organizing autonomously software processes among devices of a smart space. To do this, the authors propose the use of an intelligent system which is based on a knowledge representation of the system entities which are divided into four types of data: recipes, capabilities, rules and properties. The recipes define the contexts in which the system responds by applying rules in reaction to a particular context. The capabilities are used to define the entities participating in the system and their functionalities. Finally, the properties refer to the capacity of entities which define the presence or absence of devices, features, etc. For example, at the arrival of a request to play a song on one of the multimedia systems in a smart space, the system compares the context of the query with the contexts of basic recipes. If the conditions in the recipe are checked and there is the presence of a device with a music player (properties and capacity), a deployment policy is implemented. Similar to Ranganathan's solution approach, the Syed et al. solution is based on the orchestration. The coordinator node, the intelligent algorithms and the knowledge base are centralized on a system in each smart space.

5. Context-aware Recommender Systems

As we presented in the introduction, it is possible to impose services on users (such as in Syed or AMIGO) or recommend them, using different techniques. Using the context-aware models and recommendation algorithms to provide services or contents, Adomavicius et al. (2011) was one of the first to propose context-aware recommender system which works on integrating contextual information in a multidimensional analysis of the users' preferences (in collaborative filtering), depending of the period of the day. Other works have been done on location-based recommender system. For instance, Levandoski et al. (2012) propose a solution based on three types of location ratings (spatial rating for non-spatial item, non-spatial rating for spatial item and spatial rating for spatial item). The approach of Levandoski et al. is similar to the work of Adomavicius, where they used four-tuples or five-tuples to qualify the ratings and use multidimensional analysis techniques to compare ratings, but with an extended definition of the context.

Shin et al. (2012) propose a system that analyzes the context and history of a smart phone for classifying the different installed apps, depending on their probabilities of use. One of their conclusions is that the app transition data are one of the most important contextual informations to predict which app will be used next. Such a system can help in recommending services to the users, depending on his current smart phone usage. Similar in their approach, Huang et al. (2012) propose a system that predicts app usage based on the mobile phone context. In their case, they focused on five contextual informations: last used app, hour of the day, day of the week, location and the user profile. The two kinds of information about the time are correlated with the location of the phone. About such recommender system, in one of our latest works on the analysis of the mobile applications usage on smart phone (Gouin-Vallerand and Mezghani 2014), we concluded that the transition between application usages is distinct contextual information versus to the probability of an application uses in a same period of time. Our analysis shows also that the probabilities of transition become less obvious when the number of application usage occurrences is high (the more an apps is used, the more it is difficult to be recommended in a specific context). This conclusion also means that it can be relatively easy to recommend assistive services that are specialized, thus not used so often as compared to apps, such as the web browser or the mail application).

In conclusion, several researches have been done on the service provision in different domains of application, from smart homes to smart phones. Table 1 presents a resume of this review of the literature under specific aspects: the research settings, the types of provision approach, the type of service provided and the technologies used. Of course, our overview of the domain does not include every work on the domain of service provision. Other works, such as Ghorbel et al. (2006) on the assistance, in ubiquitous displays (Kruger et al. 2012) or in augmented reality in a smart environment (Shin et al. 2009) also include different aspects of context-aware service provision systems.

6. Review of Technologies

Different technologies exist to support the provision of services, depending on targeted devices. As discussed in the introduction, the heterogeneity

Table 1. Comparison of research projects based on the research settings, the types of provision approach, the type of service provided and the technologies used.

Works	Settings	Provision Approaches	Type of Service Provided	Technologies
EasyLiving project	Smart apartment	Orchestration	HCI and multimedia	Microsoft environment
Aura project	Office & home	Orchestration	HCI for office & home environment	Microsoft and Unix
PCOM	Smart environment	Orchestration	Components for data processing and user interactions	Java
O2S	Smart environment	Orchestration	Components	Prolog and Java
AMIGO	Smart home for PwSN	Mix of choreography and orchestration	Assistive and home services	OSGi and Java
Trumler et al.	Smart office	Choreography	Office contents and software	Native
Ranganathan et al.	Conference room	Orchestration	Multimedia and HCI	Java, STRIPS planning
Syed et al.	Smart environment	Orchestration	Assistive Services	Java and case base reasoning
Adomavicius et al.	Web apps for content provision	Recommendation	Multimedia	Web
Levandoski et al.	Smart phones	Recommendation	Mobile applications	Android OS
Shin et al.	Smart phones	Recommendation	Mobile applications	Android OS
Ke et al.	Smart phones	Recommendation	Mobile applications	Android OS

of the devices compels technological solutions that are multiplatform and usable on several types of hardware (imbedded devices, smart phones, etc.).

7. OSGi Framework

One example of such solution is the OSGI framework. The OSGi Alliance, which was better known as the Open Service Gateway initiative, is an enterprise consortium regrouping companies, such as Ericsson, Nokia, Siemens and IBM, which work together to create an open specification

(OSGi 2014) for a service-oriented software platform. This specification defines every part required for a fully functional SOA platform, such as mechanisms to manage the life cycles of plugins and services, update mechanisms for services and plugins, services' description and discovery, etc. OSGi is well known to be the software base of the Eclipse IDE plugin system, Oracle Weblogic Web Server or JBoss Application Server. However, at the beginning, the OSGi specification was primarily created to reflect the requirements of the consortium's members to be able to deploy modular software in a fast and easy way. This possibility, to deploy software rapidly and efficiently, attracted several researchers and enterprises to develop service provision systems.

As mentioned earlier, the OSGi specification is based on the Java language technology. It particularly uses the Java language introspection and class-loading mechanisms to instantiate modules and services. In OSGi, a module is called a *bundle*, which is a typical Java *JAR*-compressed file with specific data in its manifest file and specific classes. Each bundle exports or imports codebases or services, depending on their functionalities, to other bundles, creating complex functionalities, such as GUI or WebServer. In OSGi, a specific service is a bundle's class instance that offers methods (i.e. functions), which can be called by other OSGi bundles. What is interesting in the OSGi SOA model is the capability to create dynamic relations between modules, add, update or remove in runtime several bundles while reducing the impact of these actions on the quality of service. The exchange of data and utilization of services are managed by four specified layers:

- Security layer that manages the bundle validity through signature validation and hashcodes
- Module layer that manages the loading of the codebases and their executions
- Life cycle layer that manages the life cycle of bundles
- Service layer allows service exchanges between modules

Several additional specifications or specification versions exist, which provide additional capabilities or mechanisms. For instance, the *OSGi Bundle Repository* (OBR) defines mechanisms to automatically manage the bundle dependencies towards other bundles that provide required codebases or services. For instance, a bundle that would instantiate a J2EE website would require a webserver instance prior to its instantiation. OBR gives the functionalities to define such dependencies and manage the installation, starting and instantiation of a webserver bundle, prior to installing and running the J2EE website bundle. Without being as complete as the Debian Package system, OBR gives the required functionalities to reduce the overheads related to the management of the functional dependencies. Moreover, the distributed OSGi specification (OSGi v4.2) gives the specification to automatically build and instantiate Web Services version of the OSGi services, allowing remote invocation of the services by other OSGi platforms or other systems. The Apache foundation project CXF proposes a set of OSGi bundles implementing the specification. A similar specification exists for UPnP and there is also an extension to support DPWS on OSGi.

Several implementations of the OSGi platform exist with Apache Felix, Equinox OSGi and Knopflerfish being the most popular open-source implementation and Prosyst mBS one of the last remaining commercial versions of OSGi available. In the last few years, Apache Felix took an important part in the user market and proposed several pre-build bundles, such as the web server Jetty, a UPnP base driver, a Web management console, etc.

In the pervasive and ubiquitous computing research community, several works have been done that use OSGi to support service-oriented architectures. For instance, Gu et al. (2005) are among the first authors to write on the integration of OSGi to ubiquitous computing system. In their paper, Gu et al. propose the utilization of OSGi as the backbone of their context-aware system and includes an OWL description and reasoned to automatically deploy assistance services. Moreover, Vallee et al. (2005), in the context of the European project AMIGO, propose a system to dynamically create end-user services through service composition. The service composition is initiated by abstracted plans, describing environmental state/context the plans are responding to, which actions should be taken and the notification to the users. These different plan steps are matched with services in the environment through a composition manager. OSGi is the core platform of the AMIGO project, but includes other technologies, such as .Net and UPnP. Finally, the Tyche project, which is further described in the next section, uses also the OSGi platform as the base of middleware to support the autonomous configuration of smart environments based on the context, the user's profile and the taxonomy of the environment.

8. OCAP Framework

The Open Cable Application Platform (OCAP) is a software platform based on the globally executable standard Multimedia Home Platform

(MHP-GEM) to manage and deliver television services to cable television customers. Designed by Cable Television Laboratories, a research consortium cable, OCAP is both an operating system for a gateway to cable TV and a middleware for the management of services to customers.

Developed from the Java object-oriented language version of Micro Edition (J2ME), OCAP is a middleware that can be deployed across multiple types of gateways, provided there is a corresponding functional J2ME virtual machine hardware. Using the television cable, it is possible to send management requests or update software components to OCAP gateways. This middleware allows total control of the life cycle of applications for installation, starting, shutdown, dismantling and updating of applications. OCAP mainly uses the Push method to route management applications, i.e. the requests are sent from managers to gateways. Moreover, the adopted management strategy is the orchestration, i.e. the managers are the ones who manage the content of gateways.

The OCAP middleware allows to manage the entire life cycle of applications deployed on gateways to cable television through the remote system. It also offers useful tools for logging and monitoring of deployed applications. This solution has been designed for a specific type of system, television gateways and is more or less usable for service provision on other type of devices.

9. Android Operating System

The last but not the least is the Android OS that is deployed in a large range of end-user devices—smart phones, tablets, television set-top box and many other devices. The Android OS is based on a Linux Kernel and is developed by the enterprise Google. About the service provision, the Android OS includes several features that help to support the delivery of software components to users. Firstly, most of the software applications in Android are packaged within a packaging file format called *Android Application Package* (APK), which is pretty similar to the Debian package system. APK files are bundled as JAR files and include the meta-data needed to run each application in its manifest file. Android applications are usually developed in the Java language that needs to be compliant with the Dalvik VM, but can also include other code or scripts, such as HTML5 content. The APK packaging system eases the deployment and management of new applications in Android OS devices. To support even further the deployment of APK on Android devices, most of the Android OS versions (except for some versions that are not compliant with Google) include the Google Play application, which is a digital distribution system to host the different Android applications available, manage the software version and the dependencies between the different applications—the available API and the OS versions. In a way, the Google Play application is also similar to the Debian dpkg system. Thus, a service provision system on Android could use the Google Play application via its API besides using the different services to deploy new APK, manage the dependencies and the available versions.

As we present in the review of the literature, several researches have been done on context-aware recommender systems on Android devices. These works use the available contextual information provided by the operating system, the usage history and data from embedded sensors to recommend already installed APK or the other available applications.

Finally, it is also possible to deploy the OSGi framework on Android devices besides deploying the adapted bundles using the OSGi functionalities. Among others, it is possible to use Apache Felix, Knopflerfish and Prosyst mBS OSGi framework on Android OS. However, the usual Java code included in OSGi bundles needs to be compliant with the Dalvik VM and been compiled with the dx tool to produce *.dex* files. Therefore, it can be difficult or time consuming to adapt and convert existing OSGi bundles for the Android platform.

10. A Case Study: Tyche Project

The Tyche project (Gouin-Vallerand et al. 2013) is a distributed middleware that is deployed on device nodes within smart spaces, such as apartments/ residential houses and allow the deployment and management of software on environment nodes based on device capabilities and user's profile. To automatically manage the service provision, the middleware analyzes the contextual information of the environments provided by the different device nodes and sensor networks, to find which devices would fit best for hosting the services. The middleware is based on the OSGi framework and its service-oriented approach; it is then implemented in the Java language.

Before explaining in detail the middleware, a brief example illustrating a service provision would help to understand which kind of service provision and for which type of scenario the Tyche middleware is able to provide service delivery in smart environments.

Suppose an inhabitant from a smart apartment stands at the entrance of the kitchen around lunchtime. This inhabitant suffers from cognitive deficiencies that affect his time organization. Thus, to remind him to prepare his meal, his electronic agenda requests the system to provide a meal preparation assistant to the user in the kitchen area. The other information contained in the profile of the inhabitant includes—the user has a poor visual acuity and an average field of vision, moves at an average speed, has a good hand strength and workspace and prefers tactile screens to the mouse peripherals and keyboards. The meal preparation assistant doesn't need great resources except for a display to present its interface and a pointing device. In the best case, this software should be deployed in the kitchen zone.

On the other hand, the smart apartment is divided into several zones, e.g. the kitchen area, the living room, etc. Several devices and their interaction peripherals are located in these zones, especially, four devices are in the proximity of the user—a laptop at his one o'clock, a tablet at his ten o'clock, a server in a closet at his four o'clock and finally a TV with its multimedia computer behind him in the living room. Each of these devices has its own resources and different kinds of interaction peripherals. Figure 1 illustrates this example with a map of a smart apartment. In this figure, some of the

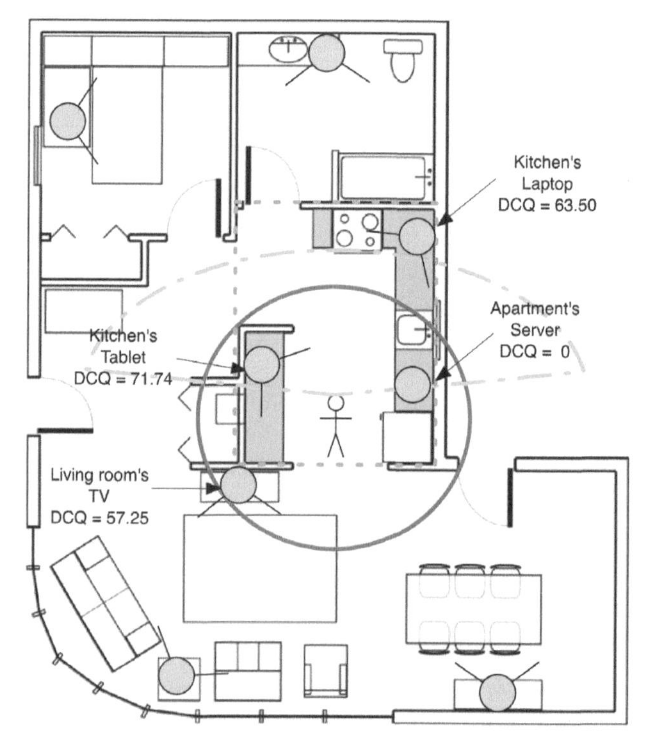

Fig. 1. A smart apartment overview illustrating the service provision scenario.

interaction modalities are shown, such as the user's visual acuity and his field of vision (the arc), the user's mobility corresponding to a walking time of two seconds or less (the circle). The kitchen zone perimeter is also indicated (the rectangle). Logically, the most suitable device in this context corresponds to the device in these three zones: the kitchen tablet. However, several other contextual information can change this logic, depending on the preference of the user or the resources' utilization of the devices.

To fulfill such a scenario, the Tyche's reasoning mechanism uses four main context's elements to deploy services toward the users: the environment device profiles, the logical configuration of the environment, the user profiles and the software profiles. Each software application that needs to be deployed or managed in the environment has its hardware, software or contextual needs. For instance, assistive applications like user-adapted agendas or a cooking assisting applications (Giroux et al. 2008) can target particular users in the environment and can require specific peripheral devices. On the other hand, users have physical capabilities and preferences about the environmental devices and the devices too have a profile with capabilities i.e. their resources, connected peripheral devices, etc. Finally, all these components are present in the smart environments at different (or not) locations and are related to contextual zones, like the kitchen, the bathroom, the living room, etc.

Therefore, the goal of Tyche service provision mechanism is to manage all the information and find the optimal organization scheme for the service to provide. Tyche functionalities are implemented in a reasoning engine—the Fuzzy Logic Organization Reasoning Engine (FLORE). The objectives of the FLORE are to match the needs of the applications to deploy in a ubiquitous environment within the context of the environment. As the environment context is a mix of quantitative and qualitative data; the Fuzzy Logic (Ross 2004) allows to do high level reasoning in a 'fuzzy' perspective, where contextual information is processed following a set of fuzzy rules. The output of the FLORE, ranging from 0 to 100, represents the Device Capabilities Quotients (DCQ), a metric value representing the viability of a device face to the software application needs. The more high is the quotient value, the more the device is closed from the optimal device target, considering the device's resources and its context.

Concerning the overall architecture of the Tyche Project and its implementation, the middleware has been built on the OSGi framework and use WebService-* standards to communicate between the environment nodes (web services and WS-Discovery). OSGi gives the support for the modularization of the ubiquitous applications and the functionalities to support the management of their life cycles. On top of the OSGi framework, we have implemented several modules that are working together to provide the service provision (Fig. 2).

The middleware includes two central components—an ontology and a reasoning engine, which work in tandem to find the best software organization for a group of applications to be deployed from the context of smart spaces. The reasoning engine, called Fuzzy Logic Organization Reasoning Engine (FLORE), uses the description logic and fuzzy logic to give an evaluation metric, the Device Capability Quotient (DCQ), of each device versus the context of the environment. The FLORE is divided into two sections—one for reasoning about the overall context of the environment (topology, user), the Organization Reasoning module, and the reasoning on the context of each device (hardware feature, location, peripheral, etc.), the Device FLORE module. These two parts of FLORE deal respectively with the macro-and micro-context (Gouin-Vallerand et al. 2012).

The ontology is instantiated and managed by the ontology management module. It is used to define the concepts of intelligent spaces (T-Box), such as devices, users, zones and their properties and relations between these concepts. It also serves to store instances of these concepts (A-Box), such as a user's profile or the profile of a particular device. The ontology was implemented using Web Ontology Language (OWL) and its concepts and instances were instantiated using the semantic platform Jena.

The implementation of the FLORE is divided into two parts which are deployed in the node coordinator and devices nodes of the smart spaces.

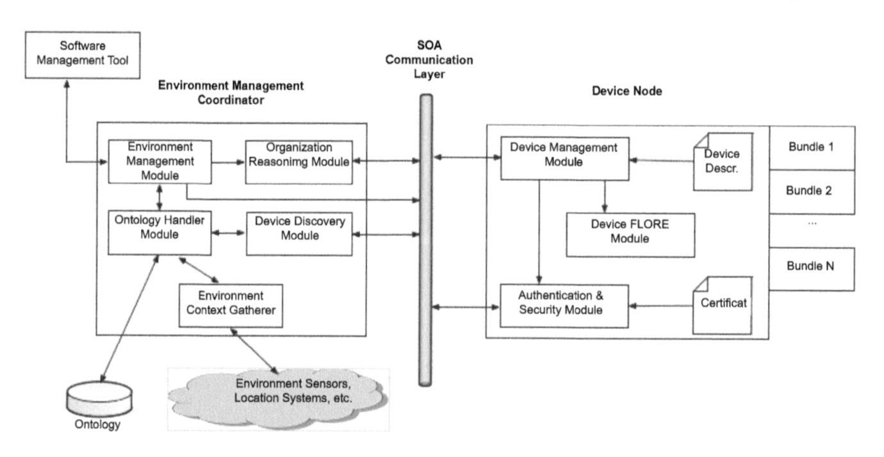

Fig. 2. The architecture of the Tyche middleware.

The FLORE-D evaluates the performance of the devices versus the needs of an application regarding the use of resources, the area where the device is located and the presence of peripheral devices (e.g. keyboard, mouse, camera, etc.). The FLORA-C evaluates at first the basic needs of an application versus the devices resources. If necessary, it assesses in a second step the user profile and context around the user versus the devices and their peripherals. Finally, he merges the DCQ from the two FLORE entities to form the final DCQ.

The FLORE algorithm has three steps according to the needs and characteristics of the applications to deploy—the general step of reasoning on the micro-context, the step about the device and user location processing and the step about processing the user profile. The first step aims at deployment of services that have resources and peripheral requirements, but do not include a deployment in a particular area of the intelligent space. In this step, the major part of the reasoning is in the micro-context of each device and is therefore done by them.

The second step is the deployment of an application in a particular area of the environment near the user. In this case, the coordinator node performs an evaluation between the devices in a specific zone and the service requirement. This evaluation uses the semantic links between the smart environments' zones to define what are the best devices to fulfill the software needs. This assessment is then returned to the devices nodes that include the assessment of the first DCQ.

The third step is the deployment of an application for a particular use by users of smart spaces. This assessment then integrates the user profile—its characteristics, preferences and capabilities and profiles of data versus the context of the environments in the calculation of the DCQ.

The sequence diagram in Fig. 3 shows the general operation of the middleware during a service provision request. The Software Management Tool initiates the service provision request; an API used by external application sends the software deployment requests. To simplify the diagram, only one device node is presented.

Thus, the Environment Manager Module receives the request from the user's tool and according to the type of request, forwards it to the FLORE coordinator module. The FLORE coordinator module:

1. Uses the Ontology Manager Module to browse the ontology for context information.

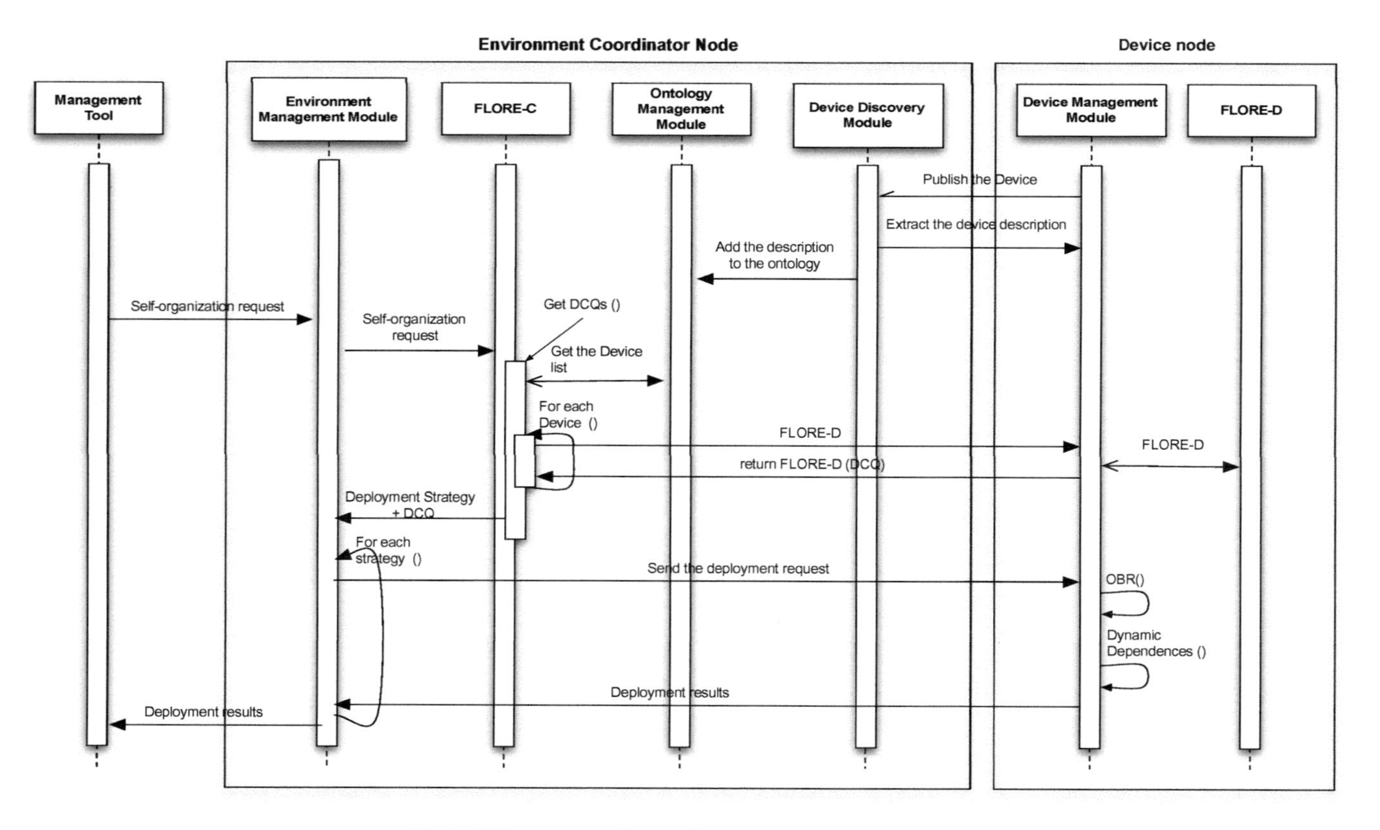

Fig. 3. Different steps during the service provision process.

2. Sends a reasoning request to the FLORE device module to reason about the micro context of the environment devices. A first DCQ related to the micro-context is returned to the FLORE coordinator module.
3. Evaluates the devices viability (DCQ) from the macro-context and merges it with the results from the micro-context (prior DCQ).
4. Depending on the set minimal threshold DCQ, selects the device with higher DCQ as the most viable device for software deployment.

For each software application contained in the service provision request, the Environment Manager Module asks the FLORE which device is the most viable one and sends a final software deployment request to the Device Management Module of the selected device.

10.1 Reasoning of the User Profile and Context

Sections of FLORE that are the most complex are those dealing with the user profile and calculating the related DCQ. The DCQ is calculated in a case where an application deployment request targets a particular user in a smart space. The DCQ related to the user profile/context is combined to the DCQ related to the device profile to form the final DCQ that is used by the FLORE to provide the service. The reasoning of the user's profile versus the context of smart spaces and applications' resources requirements is based on four types of data—its interaction capabilities, its location in space, orientation and preferences in terms of usable peripheral devices. Figure 4 represents the contextual information used by the FLORE in its organizational reasoning and in its DCQ attribution. This information is managed by an ontology written in the Ontology Web Language (OWL) and implemented in the semantic framework Jena.[1]

The FLORE bases its evaluation of user interaction capabilities with the environment on the users' interaction modalities—the senses, perceptions, motor senses and cognitive abilities (Obrenovic and Starcevic 2004). More particularly, the current version of Tyche and its FLORE uses the following interaction terms:

- *Sense–Vision*: The field of view of the users versus the computing devices and their display devices
- *Perception–Vision*: The visual acuity of the users versus the application's information on the devices' displays

[1] https://jena.apache.org/

- *Motor–Locomotion*: The user's locomotion capacity versus the devices and peripheral locations
- *Motor–Manual Interaction*: The user's physical capacities versus the peripheral devices physical needs, like hand force and hand workspace.

These modalities represent the traditional way to interact with computing devices: the vision and the sense of touch.

Of course, many other methods exist and could be evaluated as user's hearing capability versus the volume of speakers, the speech strength versus the sensitivity of a microphone or cognitive aspects, such as the language used by human-machine interfaces of applications and devices. The integration of interaction modalities in the service-provision process represents an innovation in the domain. In this work, the assessment of user interaction modalities is reduced to only two physical abilities: hand strength and opening them. The type of evaluation is categorical; users have or do not have the interaction capabilities to interact with a device.

Physical dexterity: Initially, the user's ability to interact with the peripheral devices is evaluated. This evaluation is made by comparing the physical capabilities of the user and the hardware specifications required to use the device in question. The FLORE is therefore, based on the work of Kadouche

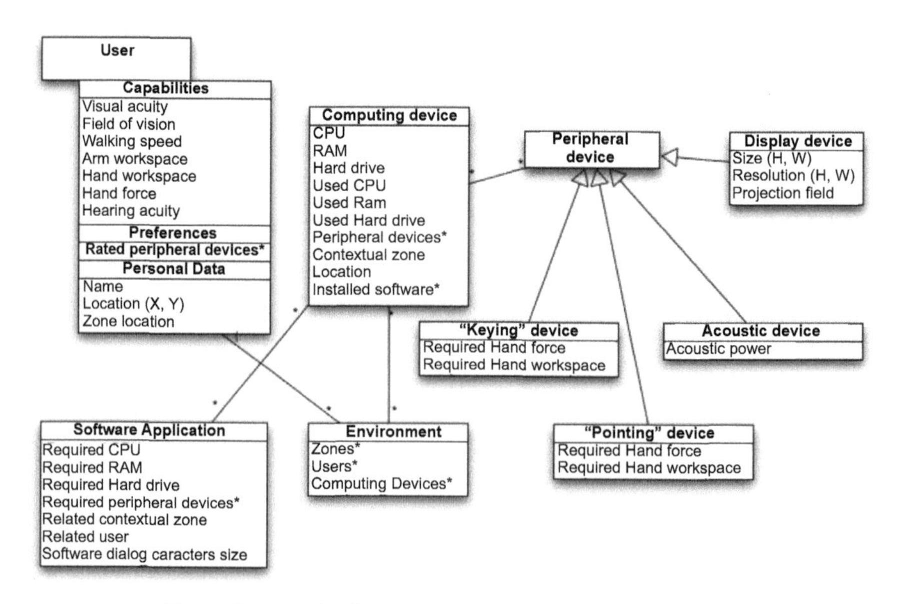

Fig. 4. Contextual information used by the Tyche middleware.

(Kadouche et al. 2009) in the field, checking the strength of the hand and the workspace user's hands i.e. the degree of opening of the hands. If this check fails, the devices directly receive DCQ worth 0 points.

Visual acuity: The FLORE uses as a second step, the user's field of view and projection field of devices to determine if the user targeted by the service provision has display devices in his field of vision. These devices would be considered as priority by the FLORE in its calculation of the DCQ. While the projection field of display devices is characterized in the contextual description by an orientation degree and a degree viewing angle, the user's field of view is defined by the user's orientation in degree and two angles of view corresponding to each eye of the user's. Thus a user can have a normal field of view for the right eye and a reduced field of vision for the left eye or vice versa. Verification of field of view is relatively simple and uses orthonormal plan changes to the user position and orientation, and then with different display devices.

Thirdly, the FLORE assesses the ability of the user to interact with applications, by calculating visual acuity ratios. This ratio value ranges between [0, 1] and is calculated from the visual acuity of the user, his/her position, the average size of dialogue characters in the service to deploy, the position of the display devices, the size of the display devices and their resolutions. The objective of this ratio is to quantify the ability of a user to read the text information on an application from its position in the environment. Obviously, the user will probably have to move to the devices where the applications are deployed to interact with them, but it can check if the user is able to recognize the applications and to have minimum information thereon.

The ratio calculation therefore, uses the average visual acuity of the user expressed with the Snellen scale (Muraoka and Ikeda 2004). This scale is widely used in optometry for quantifying visual acuity by verifying the ability of a person to read a character from five arcminutes (height and width) to a traditional distance of 20 feet, or the famous view 20/20 or 6/6 (in metres). In the case of a person with an acuity of Snellen of 10/20, it must be at a distance of 10 feet to be able to read five arcminutes of 20 feet distance, while a person with an acute 20/20 recognizes the normal distance of 20 feet. The five arcminutes correspond to scale the size of a circular arc of a five-minute corner for a circle of radius 20 feet, equivalent to a character and a height and width of 8.9 millimetres. The optotype of the user adapted to the distance of the display devices is compared with the character size of the applications on such devices. The higher the ratio the closer are the characters of the user's optotype, as when the ratio approaches

0, more will the user have difficulty to read characters on the devices from their position in the space. Ratio values are bounded [on 0, 1] to prevent large displays with a small resolution to have very high ratios and favoring some. Thus, it is checked whether the user is able to read the characters (1 ratio) or whether the user is more or less incapable (ratio of 0 or more).

User mobility: The fourth mode of interaction is assessed by the mobility of the user. This assessment is made largely by using fuzzy logic that actually contains two interrelated sub-evaluations. The first sub-evaluation is to qualify the user speed. To do this, we made a literature review of research on measures of user's movement speed. Among the different revised documents, we kept a research report on the walking speed of the urban pedestrians (Carrey 2005) and an article on the evaluation of the mobility of older persons (Abellan et al. 2009) in order to categorize and qualify the mobility of users. We have, therefore, divided the travel speeds into four types—not moving, slow moving, normal speed and finally fast walking. Each of these types of speed was associated with a normal Gaussian function that quantifies the level of membership to a correspondent walking type. The means and standard deviations were drawn from our literature review for each Gaussian function. The association of travel speeds with the user of these membership functions is used by the fuzzy logic controller FLORE-C for its calculation of the user DCQ.

The second sub-evaluation concerns the time allowed for a user to move to a device. This sub-evaluation therefore calculates the time in seconds to reach a device using the average speed of travel and the distance between the user and the device. Travel time is then described in a fuzzy set in which four membership functions are present—instantaneous travel, fast travel, slow travel and long travel. Instant travel has an average travel time between 0 seconds and 2 seconds, fast travel with 10 seconds with 2 seconds of standard deviation, slow travel with 20 seconds and 2 seconds standard deviation and finally long travel with 30 seconds and 2 seconds of standard deviation (30 seconds and higher travel time have a 100 per cent membership degree in the long travel function). Like the first undervaluation of user mobility, the degree of membership to these functions is evaluated by the fuzzy logic controller, FLORA-C.

User preferences: The user peripheral preferences for specific peripheral devices are formulated with a Likert scale to classify the environment devices. Thus, each user gives some usability preferences toward the device peripherals with values as: user likes to use the peripheral device = 1, the user is neutral face to utilization of this peripheral device = 0, the user dislikes to use the peripheral device = –1. The sum of the user's peripheral preferences for each device is calculated and used in the DCQ evaluation. These preferences can be used as a complementary tool to the physical capacity to determine the types of devices that the potential user will be able to employ. In the current version of Tyche project, five types of devices were used: keyboard, mouse, trackball, virtual keyboard and touch screen slider. Other kinds of devices can easily be added to the middleware's ontology.

In conclusion, the interaction modalities and user preferences are injected into the fuzzy logic controller that computes the user DCQ for each device of the environment that has the minimum resources to run the applications. These data are combined with the fuzzification process, using reasoning rules and then the user DCQ is calculated, using the centroid of the results of all of defuzzification. The fuzzy logic controller FLORE-C thus includes five fuzzy sets, seventeen membership functions and forty reasoning functions.

11. Discussion

The Tyche project proposes a novel middleware to support the deployment and organization of software in smart environments, such as smart homes. It uses different types of contextual information in its reasoning process and by using the Java language and OSGi framework, it copes with difficulties related to the heterogeneity of the smart space devices.

Thus, the adoption of the OSGi platform as a technological support to the implementation of Tyche was a wise choice. The dynamism provided and the reduction of software coupling enabled rapid development, easy modularity and implementation of several interchangeable solutions. The modular division of OSGi applications and lifecycle control have also played a role in the conception of the architecture of Tyche project. Also, the utilization of standard WebService-* (event, discovery, etc.) was a major element in the architectural choices and in the implementation of middleware and service provision API. Web services have allowed a multiplatform use of Tyche and expandability that other standards or protocols do not offer (Juxtapose, RMI, Jini). However, in the current version of Tyche project, the utilization of WS-* standards, the Jetty web server and Apache CXF API represents a computational load and a large utilization of the devices' memory. In particular CXF, which requires at minimum a virtual machine Java Standard Edition 1.6 (or OpenJDK 1.6) that restricts the possibilities of use on some hardware and operating system. Thus, it is impossible to run both types of Tyche node on devices with the Android operating system. However, Tyche includes bundles that use kSOAP to create accessible web services on Android.

The service-provision reasoning engine, the FLORE, of the Tyche project uses contextual information in a microscopic and macroscopic context approach. The various contextual informations on users, applications, devices and other media concepts were described, using OWL and RDF. This information is stored in an ontology in every smart space where Tyche is deployed, making it possible to make queries, data mining, inference concepts and instance, etc. The utilization of OWL/RDF allows a description of standardized bodies according to the concepts defined in the DOMUS labs, independence from hardware architecture and increased capacity for extension. Moreover, since the OWL/RDF is actually an extension of XML, it can be used in multiple systems where deployment of the Jena framework is impossible, as it was done with the FLORE-D.

Through the user descriptions and the FLORE, the Tyche project integrated the user profile to the service provision process by taking into account the abilities, preferences and user contexts. The Tyche reasoning model calculates the DCQ capabilities of a device to host applications by dealing with user profiles and applications needs. How user profiles were incorporated into the reasoning process, the intensive use of fuzzy logic for the service provision and use of DCQ represent an innovation in the community working on the service provision and software organization in smart environments. Although the model presented is limited in term of covered modalities, milestones for their use and their inclusion in service provision have been laid and the model can be refined and extended in future work.

If the Tyche middleware offers great mechanisms to support the service provision, there are still a lot of functionalities to develop and research problems to resolve. Firstly, while software applications are deployed in the environments, resources in the devices are consumed like CPU, RAM or hard drive. Thus, each request's results are dependent on the prior provision requests. Our way to deploy the various software between the environment devices is to create a priority hierarchy between the applications to deploy. When it's well used, this priority can be useful as it allows important or critical applications to have higher priorities than other applications. However, this kind of organization doesn't find the optimal organizational solution. The overall problem of the software organization is of NP complexity; the optimal solution would be find in polynomial time O(kn2) where n is the number of devices and k is the number of applications to deploy. The current algorithm is faster than the optimal solution with O(kn), but does not offer the optimal one.

Secondly, as it was presented in the review of literature, several works have been done in the last five years about the recommendation of mobile applications. As the usage of smart phones and tablets become pervasive, the recommender systems will be more and more used. The service provision approach used by the Tyche middleware is unilateral—the users do not have any way to specify which application they would prefer to assist them in their daily activities. Therefore, it would be interesting to integrate recommending algorithms that would integrate more the user preferences and also the user's history and data-mining algorithms to refine the service provision mechanisms.

About the covered interaction modalities and without going into specific cases, the FLORE reasoning process on the user profile should be extended to include modalities related to hearing, especially the cognitive aspects of the user, since the Tyche middleware aim, among others, is the assistance to people with cognitive impairment. A simple method would be to implement language processing used between devices (operating system), users and applications. On the other hand, a measure for quantifying or qualifying mnemonic capacity or functionality and/or user's general cognitive capacities versus the applications functionalities would contribute to a more adapted service provision. Such metrics exist in neuropsychology in the form of results of cognitive functions tests, such as the Mini-Mental State Examination (MMSE) (Folstein et al. 1975) or the Disability Assessment for Dementia (DAD) (Gélinas et al. 1999). However, it would be required to quantify or qualify the required cognitive capacity for the utilization of devices or software, which is not an easy task. Such work requires, *a priori*, an extensive research with neuropsychology and usability/ergonomics aspects.

Finally, the implementation of Tyche project is based on the hypothesis of deployment in closed and controlled smart environments. With the goal to expand its use as in smart urban environments, with mobile users, other interaction modalities, mechanisms of security and intensive multi-interaction, devices (such as public displays) would be required.

12. Conclusion

Ambient intelligence technologies deployed in living environments, such as homes and apartments, can assist users in their daily life activities through context-aware and personalized assistive services. With the aging of the population in the majority of developed countries, for instance 25 per cent of the Canadian population will be 65-years old and over in

2036 (icis 2011), several studies show that there will be important impacts on the societies—rise in healthcare costs, difficulty to find specialized labor, reduction in the quality of life in several senior citizens, etc. Service provision systems deployed in smart environments and on mobile devices can help to assist users with special needs, give them more autonomy and impact positively their quality of life.

Several researches has been done on the topic of the service provision in specific scenarios, such as assistance in smart homes, smart office initiatives or mobile devices. The Tyche middleware proposes a service provision system for smart environments based on the interaction modalities. This system uses the contextual information on smart environments and user profiles to find the most suitable device to host services and software that need to be provided to the environment's users. If the utilization of technologies, such as the OSGi framework, the WS-* stack, the OWL ontology and the fuzzy logic allow easy adaptation and deployment of the system to any type of smart environment, several additions to the framework are required to cover fully the users' interaction modalities, the different topologies of the smart environments, the security/privacy of users and so on. Thus, as for Tyche, we feel that a lot of work is required by the scientific community before real deployment of service provision systems on end-user devices/environments. With the idea of broader utilization of such technologies, we believe that the mobile technologies and the approach of context-aware recommender systems is a particularly interesting field of research for service provision. In that way, we are actually working on this aspect and have already published some of works on the topic (Gouin-Vallerand and Montero 2013; Gouin-Vallerand and Mezghani 2014). Also, with the increasing popularity of smart technologies, connected devices (internet) and initiatives, such as Smart Cities, we believe that service provision systems can help users in their daily activities by providing support to deliver cognitive support to the users and improve the user experience with their environment. Context-aware applications will be more and more used by the public and mechanisms to deploy and manage such software will be required and the research works presented in this chapter will be the scientific base for such systems.

Keywords: Ubiquitous Computing, Ambient Intelligence, Context-Awareness, Service Oriented Architecture

References

Abellan et al. 2009. Gait speed at usual pace as a predictor of adverse outcomes in community-dwelling older people an International Academy on Nutrition and Aging (IANA) Task Force. The Journal of Nutrition, Health amp; Aging. 13: 881–889.

Adomavicius, G. and A. Tuzhilin. 2011. Context-Aware Recommender Systems, Recommender Systems Handbook, Springer US.

Becker, C., M. Handte, G. Schiele and K. Rothermel. 2004. PCOM—a component system for pervasive computing. *In*: Proceedings of the Second IEEE Annual Conference on PerCom. 67–76.

Carrey, N. 2005. Establishing Pedestrian Walking Speeds. Technical report, Portland State University, ITE Student Chapter.

Dey, A.K., G.D. Abowd and D. Salber. 2001. A conceptual framework and a toolkit for supporting the rapid prototyping of context-aware applications, Human-Computer Interaction. 16 : 97–166.

Folstein, M.F., S.E. Folstein and P.R. McHugh. 1975. Mini-mental state—A practical method for grading the cognitive state of patients for the clinician. Journal of Psychiatric Research. 12(3): 189–198.

Gélinas, I., L. Gauthier, M. McIntyre and S. Gauthier. 1999. Development of a functional measure for persons with Alzheimers disease: the disability assessment for dementia. The American Journal of Occupational Therapy, Official Publication of the American Occupational Therapy Association. 53(5): 471–481.

Ghorbel, M., M. Mokhtari and S. Renouard. 2006. A distributed approach for assistive service provision in pervasive environment. pp. 91–100. *In*: Proc. of the 4th International Workshop on Wireless Mobile Applications and Services on WLAN Hotspots.

Giroux, S., J. Bauchet, H. Pigot, D. Lussier-Desrochers and Y. Lachappelle. 2008. Pervasive behavior tracking for cognitive assistance. International Conference on Pervasive Technologies Related to Assistive Environments PETRA. 86–93.

Gouin-Vallerand, C., P. Roy, B. Abdulrazak, S. Giroux and A.K. Dey. 2012. A macro and micro context awareness model for the provision of services in smart spaces. *In*: Proceedings of the 10th International Smart Homes and Health Telematics Conference (ICOST'12).

Gouin-Vallerand, C., B. Abdulrazak, S. Giroux and A.K. Dey. 2013. A context-aware service provision system for smart environments based on the user interaction modalities. J. Ambient Intell. Smart Environ. 5, 1 (January 2013): 47–64.

Gouin-Vallerand, C. and J.A. Montero De La Cruz. 2013. Analysis of a context-aware recommender system model for smart urban environment. *In*: Proceedings of International Conference on Advances in Mobile Computing & Multimedia (MoMM '13).

Gouin-Vallerand, C. and N. Mezghani. 2014. An analysis of the transitions between mobile application usages based on markov chains. *In*: Proceedings of the 2014 ACM International Joint Conference on Pervasive and Ubiquitous Computing; Adjunct Publication.

Gu, T., H.K. Pung and D.Q. Zhang. 2005. A service-oriented middleware for building context-aware services. J. Netw. Comput. Appl. 28: 1–18.

Hoey, J., X. Yang, E. Quintana and J. Favela. 2012. Lacasa: location and context-aware safety assistant. *In*: Proceeding of International Conference on Pervasive Computing Technologies for Healthcare.

Huang, K. et al. 2012. Predicting mobile application usage using contextual information. *In*: Proceedings of the 2012 ACM Conference on Ubiquitous Computing.

Institut Canadien d'information sur la santé (ICIS). 2011. Les soins de santé au Canada 2011: regard sur les personnes âgées et le vieillissement.
Kadouche, Rachid, Abdulrazak Bessam, Mokhtari Mounir, Giroux Sylvain and Pigot Hélène. 2009. Personalization and multi-user management in smart homes for disabled people. Journal of Smart Home. 3(1).
Kephart, J.O. and D.M. Chess. 2003. The vision of autonomic computing. IEEE Computer. 36(1): 41–50.
Krüger, A. and T. Kuflik. 2012. Ubiquitous Display Environments. Springer-Verlag, Berlin, Heidelberg.
Levandoski, J.J., M. Sarwat, A. Eldawy and M.F. Mokbel. 2012. LARS: A Location-Aware Recommender System. IEEE 28th International Conference on Data Engineering (ICDE). 450–461.
Obrenovic, Z. and D. Starcevic. 2004. Modeling Multimodal Human-Computer Interaction. ACM Computer Journal. 37: 65–72.
OSGi Alliance. 2014. OSGi Service Platform Core Specification, Release 6. Technical report, OSGi Alliance.
Muraoka, T. and H. Ikeda. 2004. Selection of display devices used at man-machine interfaces based on human factors. IEEE Transactions on Industrial Electronics. 51(2) : 501–506.
Paluska, J.M., J. Waterman, C. Terman, U. Saif, H. Pham and S. Ward. 2003. A case for goal oriented programming semantics. *In*: Workshop on System Support for Ubiquitous Computing (UbiSys'03), 5th International Conference on Ubiquitous Computing.
Preuveneers, D. et al. 2004. Towards an extensible context ontology for ambient intelligence. pp. 48–159. *In*: Ambient Intelligence, Lecture Notes in Computer Science, Vol. 3295. Springer, Berlin/Heidelberg.
Ranganathan, A. and R.H. Campbell. 2004. Autonomic pervasive computing based on planning. International Conference on Autonomic Computing. 80–87.
Ranganathan, A., C. Shankar and R. Campbell. 2005. Application polymorphism for autonomic ubiquitous computing, Multiagent Grid Syst. 1(2): 109–129.
Ratti, C. and A. Townsend. 2011. The social nexus, Scientific American, August.
Rialle, V., C. Ollivet, C. Guigui and C. Hervé. 2008. What do family caregivers of Alzheimer's disease patients desire in smart home technologies? Contrasted results of a wide survey. Methods of Information in Medicine. 47(1): 63–69.
Ross, T.J. 2004. Fuzzy Logic with Engineering Applications. Wiley, Inc.
Roy, P.C., B. Bouchard, A. Bouzouane and S. Giroux. 2007. A hybrid plan recognition model for Alzheimer's patients: Interleaved erroneous dilemma. pp. 131–137. *In*: Proc. of the 2007 IEEE/WIC/ACM International Conference on Intelligent Agent Technology, Washington, DC, USA, IEEE Computer Society. IAT '07.
Ryan, N., J. Pascoe and D. Morse. 1997. Computer Applications in Archaeology Chapter in Enhanced Reality Fieldwork: the Context-Aware Archaeological Assistant.
Shafer, S., J. Krumm, B. Brumitt, B. Meyers, M. Czerwinski and D. Robbins. 1998. The new easyliving project at microsoft research. pp. 30–31. *In*: Proc. Joint DARPA/NIST Smart Spaces Workshop.
Shin, C., L. Wonwoo, Y. Suh, H. Yoon, Y. Lee and W. Woo. 2009. CAMAR 2.0: Future direction of context-aware mobile augmented reality. International Symposium on Ubiquitous Virtual Reality. 21–24.
Shin, C., J.H. Hong and A.K. Dey. 2012. Understanding and prediction of mobile application usage for smart phones. Proceeding of UbiComp. 173–182.
Skubic, M., R.D. Guevara and M. Rantz. 2012. Testing classifiers for embedded health assessment. pp. 198–205. *In*: Donnelly, M.P., C. Paggetti, C.D. Nugent and M. Mokhtari (Eds.). ICOST. Lecture Notes in Computer Science, 940, vol. 77251: Springer.

Sousa, S.P. and D. Garlan. 2002. Aura: an architectural framework for user mobility in ubiquitous computing environments. *In*: Proceedings of the IFIP 17th World Computer Congress - TC2 Stream/3rd IEEE/IFIP Conference on Software Architecture: System Design, Development and Maintenance (WICSA 3).

Syed, A., J. Lukkien and R. Frunza. 2010. An architecture for selforganization in pervasive systems. pp. 1548–1553. *In*: Design, Automation Test in Europe Conference Exhibition (DATE).

Talwar, V., D. Milojicic, Q. Wu, C. Pu, W. Yan and G. Jung. 2005. Approaches for Service Deployment. IEEE Internet Computing. 9: 70–80.

Trumler, W., J. Petzold, F. Bagci and T. Ungerer. 2004. AMUN—Autonomic Middleware for Ubiquitous eNvironments Applied to the Smart Doorplate Project. pp. 274–275. *In*: Proceedings of the International Conference on Autonomic Computing.

Trumler, W., R. Klaus and T. Ungerer. 2006. Self-configuration via cooperative social behavior. *In*: Third International Conference on Autonomic and Trusted Computing 2006, ATC, volume 4158, Lecture Notes in Computer Science. 90–99. Springer.

Vallée, M., F. Ramparany and L. Vercouter. 2005. Flexible composition of smart device services. pp. 27–30. *In*: The 2005 International Conference on Pervasive Systems and Computing (PSC-05).

Weiser, M. 1993. Ubiquitous Computing. IEEE Computer. 26, 10 : 71–72.

6

Applying Data Mining in Smart Home

Kevin Bouchard, Frédéric Bergeron* and *Sylvain Giroux*

1. Introduction

Research on smart environments covers a broad range of topics and applications which require expertise from many disciplines (Robles and Kim 2010). Smart environments are used to improve energy efficiency and increase quality of life while being at home (Ricquebourg et al. 2006). They can also be exploited to provide more complex and more specialized services, such as the recognition of faces in the workplace to assess the mood of the employees (emotion recognition) or even to provide continuous healthcare support. Healthcare support is generally accompanied by security services and monitoring of ongoing Activity of Daily Living (ADL) (Katz et al. 1963). ADLs compose a set of common activities that a normal person is supposed to realise as autonomous (Schwartz et al. 2002; Baum and Edwards 1993). The performance in the realization of ADLs is a good indicator of the state of the resident and helps artificial intelligence to provide adequate punctual support services automatically (Nugent et al. 2007). In order to exploit the information related to ADLs, the smart home must possess an intelligent software capable of deducing the action plan

DOMUS, Universite de Sherbrooke; Sherbrooke (Quebec), Canada J1H 1H9.
Email: {Kevin.Bouchard, Frederic.Bergeron2, Sylvain.Giroux}@usherbrooke.ca
* Corresponding author

and the intended goal of the resident. That task is a well-known problem in research and is called the activity-recognition challenge (Kasteren et al. 2008; Bouchard et al. 2008).

Activity recognition is defined as the process of identifying a sequence of actions from sensors data and matching them to the corresponding model of ADL. In a smart home context, it is generally qualified as *keyhole* since the observation is performed without the knowledge of the resident (Cohen 1984). The difficulty of activity recognition in the smart home context lies in the fact that the basic actions cannot be directly observed. The sensors data are generally sparse and noisy, the source is multi-modal and preprocessing is necessary in order to use the data. Despite this, the classical approaches to this problem are mostly logic based. While there is no consensus on the classification of the activity-recognition methods, they are usually grouped under knowledge-driven and data-driven approaches. Knowledge-driven approaches were first investigated and generally assumed that a complete planned library exists to recognize the activities. That library can be encoded with first-order logic (Kautz 1991), description logic (Bouchard et al. 2007), ontology (Chen et al. 2012) or any other formalisms. It can also be described by probabilistic models, such as Hidden Markov Model (HMM) (Nguyen et al. 2003) or Bayesian network (Albrecht et al. 1998) which represents the uncertainty level in the decision process. The main problem with knowledge-driven approaches is the hard assumption they make on the planned library that has fundamental limits to the implementation of such algorithms in real deployed smart homes (Turaga et al. 2008). First, the construction of the library is performed by a human expert and it can grow very complex within a small set of a few ADLs; second, the library is static, that is, it cannot evolve over time and adapt to the profile of the residents; and finally, it is generally assumed that the library is complete (i.e. contains all possible ADLs). However, there are endless ways to perform an ADL and there is a large number of possible ADLs a human can perform in his daily life. These problems often hamper the ability of researchers to deploy real-world activity-recognition algorithms.

One avenue of solution to these limitations lies within the second family of methods—the data-driven approaches (Wyatt et al. 2005). They involve exploitation of data-mining algorithms to learn the models corresponding to the set of ADLs instead of presupposing their existence (Gu et al. 2010). Using data mining for that purpose is challenging, but the advantages are numerous. First, the library could be built automatically; second, this library could evolve with upcoming data from the sensors and adapt to the particular profile of the resident (e.g. If he is suffering from Alzheimer, his state will slowly worsen over time); third, the deployment of an assistive smart home would require less intervention from human experts. Indeed, it is next to impossible to clearly define how the sensors are bound to basic actions and thus a data-mining solution could adapt automatically. It is important to keep in mind that the configuration of a new smart home is time-consuming and costly. Finally, the same solutions could be exploited to create tools that could enable healthcare professionals to perform a closer and better monitoring of the state of the residents (Komninos and Stamou 2006). In this chapter, the reader is introduced to the main data-mining approaches and the works of researchers that applied them in the smart home context. The goal is to provide understandable material to exploit data mining for activity recognition in a smart home. This chapter can be seen as an introductory tutorial on the subject to be used as a basis for further development and research.

The remainder of this chapter is divided as follows: Section 2 discusses the general knowledge related to data mining. It defines what it is and describes the main challenges related to research and application of data mining. Section 3 introduces the reader to the decision trees. The principles are reviewed along with the main limitations for their application in smart home research. Section 4 covers the association rules on mining algorithms. Again, the basic principles are reviewed and the main works related to smart home research are reviewed. Section 5 investigates clustering for activity recognition. The main algorithms are explained and examples of their applications are provided. An opening on spatial data-mining algorithms concludes the section. Finally, Section 6 discusses the main challenges for smart home research in the next decades while concluding the chapter.

2. Data Mining Primers

Data mining is the set of methods and algorithms allowing exploration and analysis of database (Ian H. Witten and Franck 2010). It exploits tools from statistics, artificial intelligence and database management system. Data mining is used to find patterns, association, rules or trends in datasets and usually to infer knowledge on the essential parts of the information (Quinlan and Ghosh 2006). It is often seen as a subtopic of machine learning. However, machine learning is typically supervised, since the goal is to simulate the learning of known properties from *experience* (training set) in an intelligent system. Therefore, a human expert usually guides the machine in the learning phase (Barlow 1989). Within realistic situations, it is often not the case. While the two are similar in many ways, generally, in data mining, the goal is to discover previously *unknown* knowledge (Chaudhuri et al. 2011) that can then be exploited in intelligent systems and business intelligence to arrive at better decisions.

The complete process of data mining is illustrated on Fig. 1. Before beginning the cycle, it is important to understand the context and the data related to our situation; for example, what is the goal of data mining? What are the consequences of errors? Are they insignificant (e.g. marketing decision for a new product) or critical (e.g. healthcare decision support service)? Considering the nature of the data available is also important but usually for the design of data mining strategy. First of all, what types of attributes are interesting? Is there any strong association between two attributes? These are examples of questions which one should try to answer before even beginning the data-mining cycle. Once this preliminary phase is accomplished, the data mining can begin. The first step is to collect and clean the data from potentially more than one source, which can be devices, sensors, software or even websites. The goal of this step is to create the data warehouse that will be exploited for data mining. The cleaning is often not necessary. However, sometimes the data might be composed of noisy elements which are easy to remove or of attributes/objects that are known to be uninteresting for our current research. The second step consists in preparation of data in the format required by the data-mining algorithm. Sometimes in this step, the numerical values are bounded or discretized; at other times, two or more attributes can be merged together. It is also in this step that high-level knowledge (temporal or spatial relationships, etc.) can be inferred for suitable algorithms. For example, the team of Jakkula

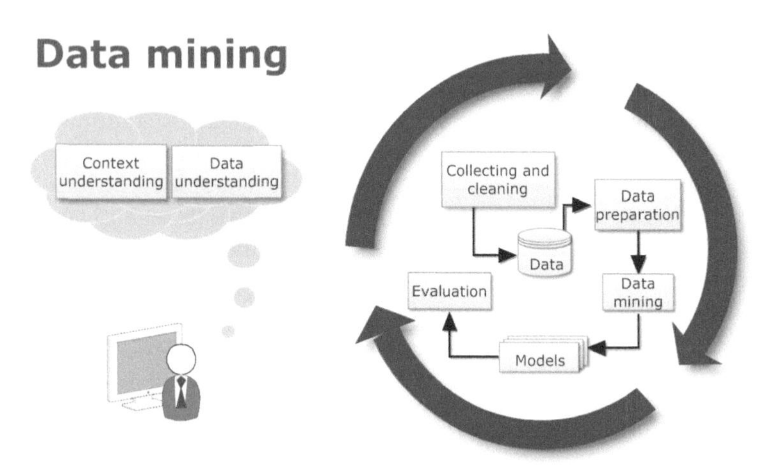

Fig. 1. The overall data-mining process.

and Cook (Jakkula and Cook 2008) exploited the temporal relationship of Allen (Allen and Hayes 1985) for association rules mining. These relations were extracted during a preparation step prior to the data-mining phase. The next step is the data mining itself. It is important to choose or design an algorithm for the context and the data. There are many algorithms to be used that we will discuss in the following sections. These algorithms are generally grouped under three main families: decision trees, association rules and clustering.

Finally, the data-mining step should result in a set of models that need to be evaluated. In a supervised context, it is usually easily done with statistical methods, such as the F-Measure, K-Statistic or the ROC curve (Ian Witten and Franck 2010). However, in an unsupervised context, it is often required to design more complex validation methods. If the evaluation is not conclusive enough, the cycle can be repeated until we are satisfied. Indeed, data mining is a method that often does not give expected results during the first time. Note that the collection and cleaning step is generally done only once, regardless of the results.

2.1 Supervised and Unsupervised Learning

Whether we talk about the data-mining method or machine learning in general, the process is usually classified under different categories (Carbonell et al. 1983). The first one is supervised learning. The method is said to be supervised since it is based on training dataset with labeled examples or classes. The signification is that the algorithm can create a model that describes each class by using the known answers in the training set. In such a situation, the idea is to generalize a function that maps the input to the output, so that it can be used to generate output for previously unseen situations. The main implication is that somehow a human expert on the subject must label the dataset. On the contrary are the unsupervised learning (Barlow 1989) works by using unlabeled examples. The idea is then to find hidden structures or associations within the dataset and generalize a model from it. The results are sometimes disappointing, whether or not hidden knowledge exists in the dataset, but also sometime very surprising as the users do not know necessarily what they are looking for. The main implication is that there is no reward signal to evaluate the potential solutions. Unsupervised learning is often more difficult to implement. Some researchers also use the name semi-supervised learning to describe their models. In that case, it usually means that the training set is partially labeled. However, it is also used to mean that unsupervised learning was applied on a training set divided into several classes by a human or an algorithm (Jakkula and J. Cook 2007).

3. Decision Trees

In the field of data mining, the algorithms are usually classified under three main categories: decision trees, association rules and clustering. The general idea behind decision trees (DTs) is to take a large set of data and find the most discriminative properties to arrive at classifying decisions. In order to do that, the training set must be labeled (i.e. each entry must have the corresponding class it belongs to). In that sense, decision trees are supervised algorithms as defined in the introduction. From that data set, the algorithm will generally go through each attribute and choose, using a heuristic, the one that best divides the instance. It will then divide the data entries using that attribute and repeat the operation for the newly created nodes. However, it is necessary to prevent overtraining. If the DT fits closely to the data, it might be impossible to classify new instances (unknown). Figure 2 shows an overfitting versus a representative model.

To prevent overtraining, a decision-tree classifier needs to have a stop condition. That condition can be a maximum branching factor, or where all attributes are used, or have a number of instances per node, etc. The classification of new instances is then simply performed by following the tree until reaching a leaf. In the next subsection, two of the most important algorithms that are exploited to construct a decision tree will be reviewed.

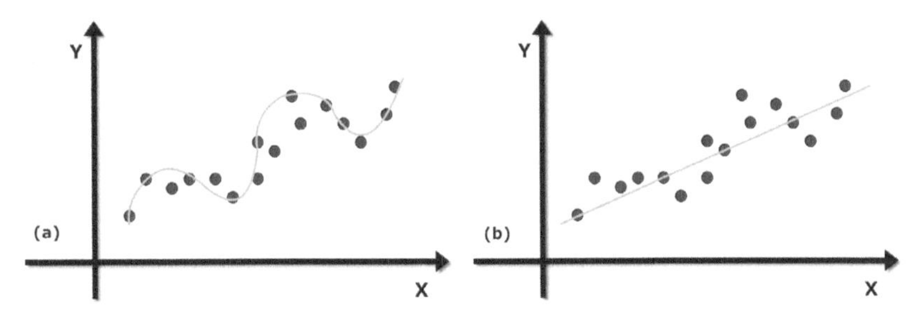

Fig. 2. (a) Overfitting the data points; (b) A more interesting and simpler model.

3.1 ID3

To illustrate the basic ideas behind decision trees, let us take a closer look into a well-known algorithm. The first algorithm that is presented is called Iterative Dichotomiser 3 or, more commonly, ID3 (Quinlan and Ghosh 2006). This precursor of the well-known C4.5 is an algorithm used to generate a decision tree from the top to the bottom without backtracking. To select the most useful attribute for classification, a criterion named the information gain based on the information theory is exploited. The information gain of a given attribute X with respect to the class attribute Y is the reduction in uncertainty about the value of Y when we know the value of X. In order to calculate the information gain, the information entropy must be known. If $E(S)$ is the information entropy of the set S and n is the number of different values of the attribute in S, and $f_s(i)$ is the frequency of the value i in the set S, then the information entropy is calculated according to the following formula:

$$E(S) = -\sum_{i=1}^{i=n} f_s(i) log_2(f_s(i)) \tag{1}$$

The entropy is always a number comprised between 0 and 1 inclusively. If all the examples are in the same class, the entropy of the population is nil. If there is the same number of positive and negative examples in binary classification, the entropy is maxed. The best attribute is selected, based on the information gain factor that is given by the following formula:

$$G(S,A) = E(S) - \sum_{i=1}^{m} f_s(A_i) E(S_{A_i}) \tag{2}$$

Where $G(S,A)$ is the gain of the set S after a split over the A attribute, m refers to the number of different values of the attribute A in S,$f_s(A_i)$ is the frequency of the items possessing A_i as i^{th} value of A in S and S_{A_i} is a subset of S. There are three requirements for the training data of ID3 algorithm—the first one is that all the training data objects must have common attributes and these attributes should be previously defined; the second requirement is that the attributes' values should be clearly indicated and a value indicating a special attribute should indicate no more than one state; the third requirement is that there must be enough test cases to distinguish valid patterns from chance occurrences. Algorithm 1 details the ID3 process:

Algorithm 1. ID3.

Input: S learning data set; the set of attributes $A = \{a_j \epsilon \{1, \ldots, p\}\}$ where p is the number of attributes remaining

If all elements in S are positive, **then**
  add $root = positive$
  Return $root$
End
If all elements in S are negative, **then**
  add $root = negative$
  Return $root$
End
If $A = \emptyset$, **then**
  add $root = negative$
  Return $root$
End

Set $a^* = arg \max_{a \epsilon A} gain(S, a)$
Set $root = a^*$
For all values v_i of a^*
  add a branch to $root$ corresponding to v_i
  Create $S_{a^*=v_i} \subset S$
  If $S_{a^*=v_i} = \emptyset$, **then**
    put a leaf with the most common value of the class among S at the extremity of this branch
  Else
    put ID3($S_{a^*=v_i}$,$A - \{a^*\}$) at the extremity of this branch
  End
End

ID3 possesses the advantage that it is fast and it builds short trees. Nevertheless, if a small sample is tested, only one attribute at a time is tested for making a decision and classifying continuous data may be computationally expensive. As any other DT algorithm, data may be overfitted or overclassified by ID3. The classes created by ID3 are inductive, meaning that given a small set of training instances, the specific classes created by ID3 are expected to work in all future instances. A limitation of ID3 is that the distribution of the unknown conditions must be the same as the test cases and the induced classes cannot be proven to work in every case since they may classify an infinite number of instances.

Example of Construction of a DT

To show the main characteristics of the construction of a decision tree with ID3, consider the example dataset found on Table 1.

From this dataset S, the overall entropy is:

Table 1. Example dataset.

Meal	Filling	Size	Pattern	Class
Poutine	Ketchup	Small	Filled	BBQ
Hot-dog	Mustard	Small	Filled	BBQ
Hot dog	Ketchup	Small	Striped	Oven
Pizza	Mustard	Big	Striped	BBQ
Poutine	Mustard	Big	Striped	BBQ
Hot-dog	Ketchup	Medium	Filled	BBQ
Pizza	Mayonnaise	Big	Striped	Oven
Pizza	Ketchup	Medium	Striped	Oven

$$E(S) = \frac{5}{8}log_2\left(\frac{5}{8}\right) + \frac{3}{8}log_2\left(\frac{3}{8}\right) \approx 0.9544$$

To construct the tree, the information gain for each attribute needs to be calculated next. For example, the calculation for the attribute *Filling* is:

$$G(S, Filling) = E(S) - \left(\frac{4}{8}\mathrm{E}(2,2) + \frac{3}{8}\mathrm{E}(3,0) + \frac{1}{8}\mathrm{E}(0,1)\right)$$

$$G(S, Filling) = E(S) - \left(\frac{4}{8} * 1 + \frac{3}{8} * 0 + \frac{1}{8} * 0\right)$$

$$G(S, Filling) = E(S) - 0.5 \approx 0.4544$$

Note that there are three entropy calculations made for each possible value of the attribute. For instance, $\frac{4}{8}\mathrm{E}(2,2)$ is part for *Ketchup* and 4 out of 8 are octagonal. Here 2,2 means that two of the *Ketchup* data entries are for BBQ and two are for the *Oven*. The gain of the three others attributes is:

$$G(S, Meal) = E(S) - \left(\frac{2}{8}\mathrm{E}(2,0) + \frac{3}{8}\mathrm{E}(2,1) + \frac{3}{8}\mathrm{E}(1,2)\right) \approx 0.2657$$

$$G(S, Size) = E(S) - \left(\frac{3}{8}\mathrm{E}(2,1) + \frac{2}{8}\mathrm{E}(1,1) + \frac{3}{8}\mathrm{E}(2,1)\right) \approx 0.0157$$

$$G(S, Pattern) = E(S) - \left(\frac{3}{8}\mathrm{E}(3,0) + \frac{5}{8}\mathrm{E}(2,3)\right) \approx 0.3476$$

As it can be seen, the *Filling* gives the highest information gain; thus it is chosen as the root of the DT. The tree has three branches after this first iteration as shown on Fig. 3.

Now, only the *Ketchup* branch does not enable to clearly classify the population of the training set. The entropy of the *Ketchup* subset (S_{ket}) must be calculated followed by the information gain for the remaining attributes. In that case, the calculation is:

$$G(S_{ket}, Meal) = E(S_{ket}) - \left(\frac{1}{4}\mathrm{E}(1{,}0) + \frac{2}{4}\mathrm{E}(1{,}1) + \frac{1}{4}\mathrm{E}(0{,}1)\right) = 0.5$$

$$G(S_{ket}, Size) = E(S_{ket}) - \left(\frac{2}{4}\mathrm{E}(1{,}1) + \frac{2}{4}\mathrm{E}(1{,}1)\right) = 0$$

$$G(S_{ket}, Pattern) = E(S_{ket}) - \left(\frac{2}{4}\mathrm{E}(2{,}0) + \frac{2}{4}\mathrm{E}(0{,}2)\right) = 1$$

As it can be seen, the *Pattern* value gives a maximal information gain for the subset S_{ket} and thus it is chosen to construct the DT. The final decision tree is illustrated by the Fig. 4.

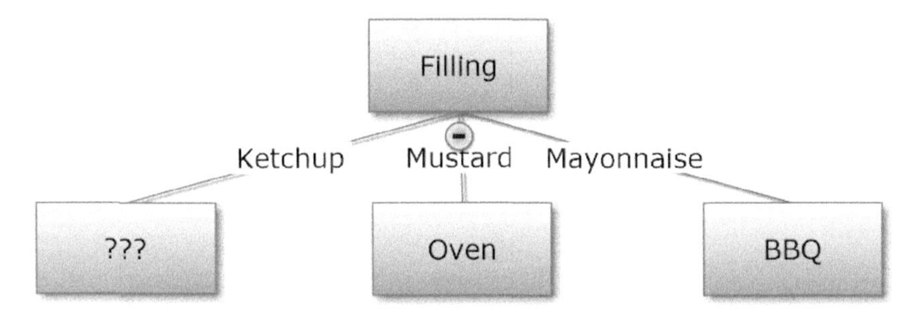

Fig. 3. The DT after one iteration.

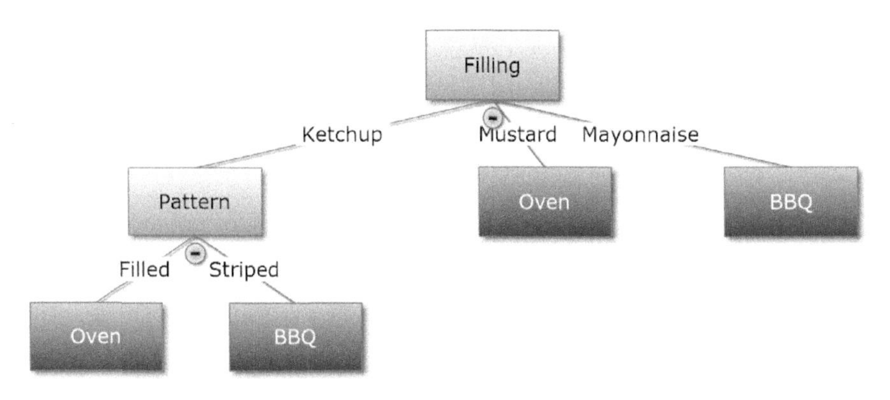

Fig. 4. The resulting decision tree for the example dataset.

3.1.1 Variants of the decision tree

Over time, many variants of the basic decision tree have been developed. However, the basic idea is always similar to what was shown until now. In this sub-section, we describe one of the interesting ones for advanced usage. Section 3.1 presented the ID3 algorithm as a tree that overfit the data. One way to avoid overfitting is by providing more training data. But, acquiring enough data is not always possible due to time or money constraints. Another way to prevent overtraining is to only consider a subset of the available information when training the model. Of course, this subset must be chosen wisely and in a certain way that it does not miss out on important things.

This idea of considering only a part of the data led to the tree called Random Tree (RT). The RT works exactly like ID3, with the simple difference that at each node, only subset of the attributes is considered for information gain. The subset is chosen randomly at each node and the entropy computed for attributes in this subset. This means the split of a given node might not be optimal, thus reducing the chances of overfitting. Indeed, ID3 supposes that future data will always have the same distribution over its attributes to classify. By not using the best split attribute, the importance of this supposition can be weakened in the final accuracy.

The key point of the random tree is to choose the attributes that will be used by this line of the previous pseudo-code of ID3:

$$\textbf{Set } a^* = arg\, \max_{a \in A} gain(S, a) \tag{3}$$

There are many ways to do so. Let's call k the number of attributes chosen at a given node. A is again the set of all attributes B and is the subset. One common way is to randomly pick k times an attribute a from A without repetition. At each step, each remaining a has the probability $\frac{1}{|A|-|B|}$ of being chosen. The last remaining point is to choose the value for k. Again, one common way is to have a varying k that will adjust its value to the number of remaining attributes as in:

$$k = \text{int}(\log_2(|A| + 1)) \tag{4}$$

From this, a simple tree that is not optimal but that should still be doing well with a good value of k and true randomness is obtained. With true randomness, there is no way of predicting what tree will be constructed. Thus, there are many possible trees for the same data. Is one really better than the other? If so, could we still consider the other when classifying? If they could work together, might the accuracy improve? These questions, among others, led to the creation of the Random Forest (RF) algorithm. The idea is simple—train many RT, let's say N, and make them vote for the class they think best represents a given instance. The number of RT is then another hyper-parameter that must be optimized along with k and the ones present in ID3. A large N might overfit the data, while a small N might not improve over a single RT.

3.1.2 The work of Stankovski

Stankovski is one of the researchers to apply the decision trees algorithm in a smart home context (Stankovski 2006). As for any DT-based system, the first step consists of building a supervised dataset. In that case, the dataset contains the whereabouts of a person; interactions with appliances, duration, etc. The DT was created; so the usual rules describing the normal setting leading to a particular event in the smart home could be known. The events occurring outside the normal setting were considered as abnormal behaviors and in such a case assistance could be triggered (alarm, message, etc.). To create the training dataset, heavy human expert intervention was required. Subsequently the observations are gathered while the expert needs to specify two more data fields. For each record of observation, he needs to assign an activity and mark which records are normal (usual).

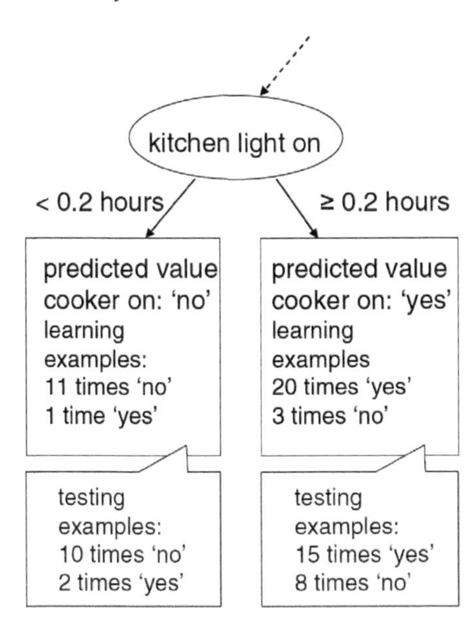

Fig. 5. A part of a decision tree induced in (Stankovski 2006).

The construction of the decision tree is done with ID3. Figure 5 shows a part of the decision tree built by Stankovski from a dataset of 35 examples.

3.1.3 The work of Bao and Intille

Bao and Intille (2004) also worked on supervised learning methods for activity recognition. For this purpose, they collected datasets from subjects wearing five biaxial accelerometers. These subjects performed many activities and 20 of them were studied for this work. The authors extracted four features from the acceleration data—the mean, the energy, the entropy and the correlation. They tried various algorithms on these datasets, including a C4.5 decision tree (direct successor of ID3). They trained a decision table, a nearest neighbor, a naïve Bayes and a C4.5 classifier using two protocols. Under the first one, each classifier was trained on each subject's activity sequence data and tested on a second dataset where they performed a list of tasks on a sheet. Under the second protocol, they trained the classifiers on all dataset for all subjects, except one. The classifiers were tested on the left out subject's datasets. This protocol was repeated for all the 20 subjects. As you can see in Table 2, they have found that the decision tree outperformed all the other classifiers. In fact, only the nearest neighbor gave close results.

Table 2. Results of the classifiers from Bao and Intille (2004).

Classifier	User-specific Training	Leave One Out
Decision Table	36.32 ± 14.501	46.75 ± 9.296
Nearest Neighbor	69.21 ± 6.822	82.70 ± 6.416
Naïve Bayes	34.94 ± 5.818	52.35 ± 1.690
C4.5	71.58 ± 7.438	84.26 ± 5.178

3.1.4 The work of Ravi et al.

Ravi et al. (2005) worked on the same challenge as Bao and Intille (2004)—activity recognition using accelerometer data. However, they wanted to use only one triaxial accelerometer worn near the pelvic region. They also formulated their problem as one of classification. Similarly, they tested well-known base classifiers, though their focus was on meta classifiers. The principle behind these meta classifiers is usually to exploit knowledge on the problem (properties) or combine multiple classifiers to improve the results. They are divided among two families. The first one is voting. It includes methods such as Boosting (Meir and Rätsch 2003) and Bagging (Quinlan 1996) and both can be used in combination to classic classifiers to improve their performance. Boosting basically applies a single algorithm repeatedly and combines the hypothesis learned each time by voting. The voting is done by weighting each training example, depending on either it was correctly or incorrectly classified during a learning iteration. Bagging is similar but works by training each classifier on a random redistribution of the dataset (it also uses only one algorithm). The second family is named Stacking and includes Meta Decision Trees (MDT) and Ordinary Decision Trees (ODT) (Todorovski and Džeroski 2003). MDT learns a meta-level decision tree whose leaves consist of each of the base classifiers. ODT instead specify the class of the given test instance at the leaves level. Therefore, MDT is an improvement on ODT which specifies the classifier to use to optimally classify an instance.

In their work, Ravi et al. (2005) have shown that the performance of decision trees in activity recognition context can be improved with meta parameters. Nevertheless, the difference in performance is often not significant as is in the case of simple decision tree classification results (97.29-98.53-77.95-57.00) versus Boosted decision tree (98.15-98.35-77.95-57.00) and Bagged decision tree (97.29-95.22-78.82-63.33). However, they found that MDT and ODT both produce higher classification accuracy for all the four settings they tested.

The bottom line

There are many advantages in use of decision trees. First, they create models that are easy to understand and use from a human perspective. They are also very robust to missing data and noise (which are present in smart homes). Furthermore, the classification (not the learning phase) is very fast and therefore makes them well suited for online activity recognition in smart homes. There are many models of decision trees that have been exploited in activity-recognition researches, such as the famous ID3 (Bao and Intille 2004) or Meta Decision Tree (MDT) (Ravi et al. 2005). There are two types of applications of DTs in the literature. They are often used to perform low granularity AR from a very specific type of information. These works focus on the technological platform rather than on the algorithm and mostly demonstrate the feasibility of their idea. For example, Ravi et al. (2005) wanted to recognize ADLs from only one simple accelerometer worn by a subject at the belt level. The other type used decision trees in combination with another approach of AR (usually clustering, but it can also be a classical artificial intelligence approach). The DT then acts as a post filtering classifier (Gaddam et al. 2007).

The main problem with DTs is that they require a large set of labeled data to perform well. If there is not enough training data, the selected attributes might mislead and the resulting classification performance is poor. Figure 6 shows a simple yet stunning example of what can happen if the training set is too small. DTs also do not really support data evolution; that is learning must be redone if the data change too much (new attributes, new type of values, new number range, etc.). Finally, the last but probably the most important limitation of AR is their weakness to distinguish a large number of classes within a dataset.

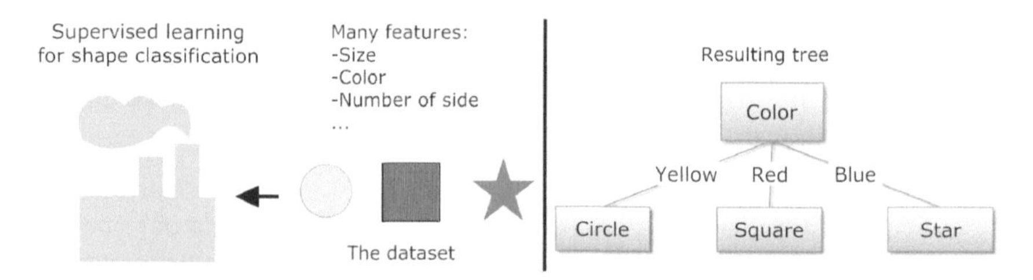

Fig. 6. A three-example dataset for shape classification resulting in a strange DT based on the color.

Association rules

Association rules mining are often confused with decision trees since the latest can always be represented by a set of rules. However, in most situations, rules are different than trees. For example, a large tree can often be represented by a smaller equivalent set of rules instead of the exhaustive decision nodes of the tree. Additionally, DTs try to split all classes while association rules mining considers one class at the time. Finally, association rules do not require labeled dataset. They are generally considered as fully unsupervised data mining algorithms. Nevertheless, in the smart home context, they are often exploited as semi-supervised algorithms.

An association rule is a rule of the form *condition => consequence* that aims to find underlying relations between the data. For example, we have a dataset comprising transactions made at a store by customers. A rule such as *if Saturday and Beers => Potato chips* could be discovered. That rule would mean that *very often*, when it is Saturday and someone buy beer, that person will also buy potato chips. Association rules mining algorithms define the terms *very often* with two attributes named the support and the confidence. The first one defines the minimum frequency of both the left and right part of the rule. For example, suppose we have the item set {{A}, {B}, {AB}, {BA},

{B}, {AB}, {AB}}, the support of AB would be *Support* ({*AB*}) = $\frac{3}{7}$ ≈ 43%. The second, the confidence, is the probability threshold of the right part being true if the left part is validated:

$$Confidence(X => Y) = p(Y|X) = \frac{p(X \cup Y)}{P(X)} = \frac{Support\ (\{AB\})}{Support\ (\{A\})} \quad (5)$$

Apriori

The main principles behind the various association rules mining algorithms are very similar. To understand those principles, one must take a closer look at the most renowned and perhaps the most important algorithm of the family. It is named Apriori and was introduced by Agrawal and Srikant (1994). It relies upon two principles—the first is research for frequent *k*-itemsets whose support is higher than a fixed minimum support; the second consists of building the association rules from the found frequent *k*-itemsets. A rule is retained only if its confidence is higher than a fixed minimum confidence.

The Algorithm 2 shows the phase one of the Apriori algorithm

Algorithm 2. Apriori, first phase.

Input: S learning data set; minimum support (σ) and confidence thresholds

Output: Set of frequent itemsets

Fetch the item sets that whose > $\sigma \rightarrow L_1$
Set $k = 1$
Repeat
  Increase k
  From L_{k-1} finds C_k the set of frequent itemsets candidates comprising k items
  Set $C_k = L_{k-1} \times L_{k-1}$
  Set $L_k = 0$
  For all $e \in C_k$ **do**
    If $Support(e) > \sigma$ **Then**
      Add e to L_k
    End
  End
Until $L_k \neq \emptyset$

Generalized sequential pattern

Another interesting algorithm that was also introduced by the team of Agrawal (Srikant and Agrawal 1996) is Generalized Sequential Pattern (GSP). This algorithm relies on the same foundation as Apriori but was developed to work precisely on data sets in sequence of transactions instead of simple transactional data. It means that the algorithm does not only take into account the presence of items together, but also the sequential ordering. Another peculiarity of GSP is its capability to exploit a taxonomy by encoding it in the data set. Let's look at an example from the original paper of Agrawal. Consider the sequence <(Foundation, Ringworld) (Second Foundation)> and the taxonomy shown in Fig. 7. To exploit the said taxonomy, all that is required to do is to integrate it directly in the data set: <(Foundation, Ringworld, Asimov, Nirven, Science Fiction) (Second Foundation, Asimov, Science Fiction)>. It is also possible to optimize the encoding in order to avoid the explosion of data (Srikant and Agrawal 1996).

Another interesting part of the GSP algorithm is pruning which is done directly on the candidate itemsets by introducing the concept of contiguous subsequence. The idea is to suppress the candidates who possess a (*k-1*)-sequence contiguous with a support smaller than the fixed minimum support. A subsequence contiguous *c* of *s* is a sequence for which one of the three criteria is true:

1. *c* derivates from *s* by rejecting either s_1 or s_k
2. *c* derivates from *s* by rejecting an item from a s_i which possess at least two items
3. *c* is a contiguous subsequence of *c′* which is a contiguous subsequence of *s*

For example, consider the set *s*=<(1, 2) (3, 4) (5) (6)>. The subsequence <(2) (3, 4) (5)>, <(1, 2) (3) (5) (6)> and <(3) (5)> are all contiguous subsequence of *s*. However, <(1, 2) (3, 4) (6)> and <(1) (5) (6)> are not. Now let's look at dataset to demonstrate the pruning work within the GSP algorithm. Consider the seed set consisting of six frequent 3-sequences:

1. <(1, 2) (3)>	3. <(1) (3, 4)>	5. <(2) (3, 4)>
2. <(1, 2) (4)>	4. <(1, 3) (5)>	6. <(2) (3) (5)>

The junction step of the algorithm would lead to obtain these two frequent 4-sequence considering a support of 100 per cent: <(1,2)(3,4)> and <(1,2)

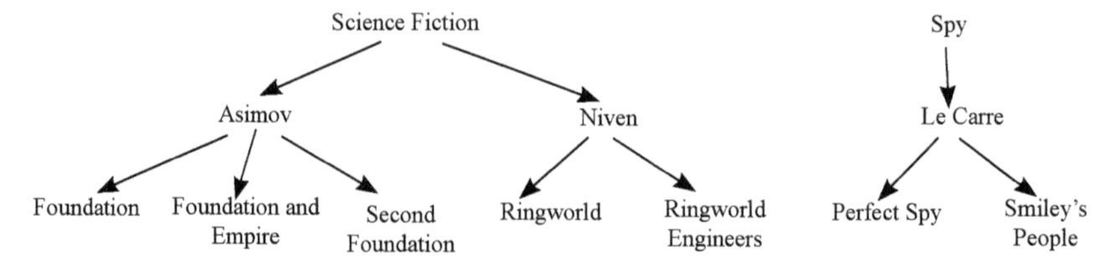

Fig. 7. Example of taxonomy from Srikant and Agrawal (1996).

(3)(5)>. The second sequence, <(1,2)(3)(5)>, would be abandoned during pruning because subsequence <(1)(3)(5)> is not part of L_3 (for GSP, the fourth sequence is not equivalent to <(1)(3)(5)>). In fact, this sequence is contiguous since the criterion number two is true for it. The next subsection will describe a complete smart home solution exploiting rules mining for activity recognition and activity prediction.

3.1.5 The work of Jakkula and Cook

Jakkula and Cook developed a renowned approach of activity discovery for smart home which is based on association rules mining. Their system was built in a multi-agent (Weiss 2000) architecture where the agents perceive directly the state of the environment from sensor's output raw data.

They collected temporal information constructed from Allen's intervals calculus presented in Allen and Hayes (1985). Their goal was to process raw data to discover frequent sequential patterns, to discover the temporal links existing between frequent events. For example, if recorded data tends to demonstrate that every time *Take Tea* happens, the kettle is activated soon after, the recognition system will infer a temporal rule from Allen's thirteen relations (*Boil Water* after *Take Tea*). Suppose that a lot of training data are available. Jakkula and Cook's model works as follows: First, the temporal intervals are found using the timestamp of events and the on/off state of binary sensors. The algorithm that associates these intervals to one of Allens' relations is illustrated below (Algorithm 3):

Algorithm 3. Temporal Interval Analyzer (Jakkula and J. Cook 2007).

Input: $E = \{\text{set of events}\}$

Output: Set of Allen's relation

```
Repeat
    While (E_i && E_{i+1})
        Find pair ON/OFF events in data to determine temporal range
        Read next event and find temporal range
        Associate Allen's relation between events
        Increment Event pointer
    End
Until end of input
```

The algorithm loops until all the pairs of events are compared. Between each pair, it establishes the Allen's relationship from the beginning to end markers of both the events.

The second step in their model is to identify frequent activities or events that occur during a day to establish a reduced set of activities. This step is mandatory because there are too much data from smart home sensors and many potential anomalies are just noises that should be ignored. They accomplish this task using the Apriori algorithm (Jakkula and Cook 2007) that was described previously. In their work, Jakkula and Cook not only demonstrated that temporal relationships provide insights into patterns of resident behaviors, but also enhance the construction of other smart home assistance algorithms. To do so, they calculated the probability that a certain hypothetic event occurs or not, given the observed occurrence of other events temporally related. The frequency of nine relationships out of thirteen they determined that could affect anomaly detection: *before, contains, overlaps, meets, starts, started-by, finishes, finished-by* and *equals*. The formula to calculate the evidence of the occurrence of an event X is given by the observation of other events (such as Y) that are temporally related (from previous learning phase). Equation 6 below allows such calculus:

$$\begin{aligned} P(X|Y) = {} & |After(Y,X)| + |During(Y,X)| + |OverlappedBy(Y,X)| \\ & + |MetBy(Y,X)| + |Starts(Y,X)| + |StartedBy(Y,X)| \\ & + |Equals(Y,X)| \;/\; |Y| \end{aligned} \tag{6}$$

The equation gives the likelihood of X considering Y. To combine evidence of X from multiple events that are in temporal relationship with X, the equation needs to be improved. Consider the events Z, Y that had been observed in this order. The prediction of X is given by the formula $Prediction_x = P(X)$ that is calculated as follows (7):

$$\begin{aligned} P(X|Z \cup Y) &= \frac{P(X \cap (Z \cup Y))}{P(Z \cup Y)} = P(X \cap Z) \cup \frac{P(X \cap Y)}{P(Z)} + P(Y) - P(Z \cap Y) \\ &= P(X|Z).P(Z) + P(X|Y).\frac{P(Y)}{P(Z)} + P(Y) - P(Z \cap Y) \end{aligned} \tag{7}$$

From the formula, anomalies can be detected and predictions can be made. If an event X is a probability approaching 1, then it is considered as most likely to occur. On the other hand, if its probability is close to 0, it will be considered as an unusual event and will be ignored from further predictions. The final step is to use an enhanced version of *Active LeZi* (ALZ) (Gopalratnam and Cook 2003) algorithm for the prediction by adding these discovered temporal rules as input data. This predictor is sequential and employs incremental parsing and uses Markov models. It should be noted that improved *ALZ* could be used for anomaly detection. This could be done by using the prediction as input in an anomaly's detection algorithm and by comparing prediction sequence with observations. Thus, if the new observation does not correspond to the expected event, an assisting sequence could be triggered. The add-on to the *Active LeZi* is shown below (Algorithm 4):

Algorithm 4. Temporal Rules Enhanced prediction.

Input: Output of ALZ a, Best rules r, Temporal dataset

```
While (a! = null)
    Repeat
        Set r1 to the first event in the first best rule
        If (r1 == a) Then
            If (Relation! = "After"), then
                Calculate evidence
                If (Evidence > (Mean + 2 Std.Dev.`
                    Make event in the best rule as next
                Else
                    *Get next predicted event and lool
                    temporal relations database based
                If the relation is after again, then
                    Go to * Until no more after relatic
                    If high, then predict;
                   Else Calculate evidence and if hi
                   relation; continue.
                End
            End
        End
    Until end of rules
End While
```

Following the creation of this algorithm, they conducted experiments that can be seen in Table 3. It shows the accuracy of the observed prediction performance on real and synthetic data sets. There is performance improvement in the prediction of activities of the resident of the intelligent environment. The main reason for the important error rate is the small amount of data in the datasets used for learning. The search for knowledge-based temporal rules is a new area of research in smart homes and should be further explored in the future. Note that the use of temporal relationships provides a unique new approach for prediction.

Table 3. Comparison of ALZ prediction with and without temporal rules.

Datasets	Percentage Accuracy	Percentage Error
Real (without rules)	55	45
Real (with rules)	56	44
Synthetic (without rules)	64	36
Synthetic (with rules)	69	31

3.1.6 The work of Bouchard et al.

In the same line of idea of Jakkula and Cook (2008), Bouchard et al. (2013) worked with association rules mining for activity recognition in a smart home. To do so, they exploited the topological relationships that exist between entities present in the smart environment. It was done by using the framework of Egenhofer and Franzosa (1991) which defines the relation between two entities e_1 et e_2 with the formal intersection structure between their interior (°) and boundary (∂) points: $< \partial e_1 \cap \partial e_2, e_1^{\circ} \cap e_2^{\circ}, \partial e_1 \cap e_2^{\circ}, e_1^{\circ} \cap \partial e_2 >$. By using the simple invariant empty property of sets, there are sixteen possible relation types; however, only eight exist for physical regions without holes.

In their model, activities are defined by a set of constraints K such that:

$$K = \{T(e_1, e_2) | e_1, e_2 \subseteq O \times O \cup R \times A\} \quad (8)$$

where T is a topological relation, O is a physical object, R is the resident and A is a logical area of the smart home. The recognition process includes evaluating the plausibility of each ADLs in the knowledge base from the constraint observations made in the environment. The plausibility is calculated by using the neighborhood graph of the topological relationships. Considering that the *similarity* function returns the percentage of similarity from 0 to 100 per cent between observed topological relationships and those known to define an activity, the scoring of an activity ($a_{\delta,i}$) for a single iteration i is:

$$a_{\delta,i} = \sum_{n=a_{t,0}}^{a_{t,i} \in a_T} \sum_{m=l_0}^{l_j \in L} \varphi * similarity(n, m) \quad (9)$$

where a_T is the set of topological relationships defining the activity a. It is the same calculation for the topological relationships implying the resident and a smart home zone. The next step in the algorithm is to choose the most plausible activity that is ongoing. In other words, it has to choose which

ADL best explains the observations made until the current iteration. The plausibility of an activity after i_c iterations is calculated by the function below:

$$plausibility\ (a) = \sum_{i=0}^{i_c} a_{\delta,i} * \phi^{i_c - i} \tag{10}$$

That is, the plausibility of a is the sum of all the gained points modulated by an inverted exponential function. The constant parameter $\phi \in (0, 1)$ modulates the speed at which the function tends to 0. The bigger it is, the longer is the impact of iteration's score. The last step is to normalize the points gained by the activities with equation:

$$Normalized\ ADL = \bigcup_{i=0}^{a_i \in ADL} \frac{score(a_i)}{\sum_{j=0}^{x_j \in ADL} score(x_j)} \tag{11}$$

The ADL with the highest score is the one selected as currently being realized. They exploited GSP and Aprori to automatically build an activity library for their recognition algorithm. To collect their datasets, they used a real subject that performed four different activities three times each. They collected more than 350,000 lines of data which they used with both GSP and Apriori. From the learned spatial rules, they were able to recognize 100 per cent of the activities in real time in the smart home.

The Bottom Line

As you can see, association rules mining approaches are very interesting and more general than DT. Due to their inherent working, they are perfectly adapted to learn logical rules about activity of daily living and be exploited for AR. Despite this, association rules mining usually results in an important number of trivial and non-interesting rules. That is, a human usually needs to check all the extracted rules in order to find the few that could be exploited. This supplementary step often requires a lot of effort and time and constitutes one of the major limitations. Additionally, the collected dataset is not always adapted to such algorithms. They work well on logical information, such as spatial or temporal relationship, but are less suited to deal with raw data from sensors. Thus, it is often necessary to conceive an ad hoc method to transform the data into an appropriate form. Finally, the method is not working well for rare items. Due to the high dimensionality of our data, frequent patterns might not be that frequent in real contexts.

Clustering

To address the issues that exist with DTs and association rule mining, many researchers aim to exploit completely unsupervised learning. Clustering could be a good solution since it can extract similar data automatically from unlabeled data. The idea behind this type of algorithm is simple. The goal is to find *clusters* in the dataset that could separate the records into a number of similar classes. A cluster is, in that context, a set of similar objects, where similarity is defined by some distance measure. The goal of the distance measure is to obtain clusters with a high intraclass similarity and a low interclass similarity. The distance measure should respect these four properties:

1. $d(x,y) \geq 0$
2. $d(x,y) = 0 \; iff \; x = y$
3. $d(x,y) = d(y,x)$
4. $d(x,z) \leq d(x,y) + d(y,z)$

Among the popular known distances, here are respectively the Euclidian distance, the Manhattan distance and the Minkowski distance:

$$d(x,y) = \sqrt{\sum_{i=1}^{n}(x_i - y_i)^2} \qquad d(x,y) = \sum_{i=1}^{n}|x_i - y_i| \qquad d(x,y) = \sqrt[q]{\sum_{i=1}^{n}|x_i - y_i|^q}$$

Euclidian *Manhattan* *Minkowski*

The clustering problem is a difficult challenge because the attributes (or features) and their values that differentiate one cluster from another are not known. There is no data example to tell what features differentiate objects that belong to different clusters and as the size of the dataset increases, the number of clusters, as well as the number and type of differentiating factors, might change. Moreover, there is no guide to indicate what constitutes a cluster and the success of the clustering algorithms is influenced by the presence of noise in the data, missing data and outliers.

K-Means

The most important clustering algorithm is without a doubt the *K-Means* (Selim and Ismail 1984). The goal of this algorithm is to split a dataset into k clusters where the value of k is selected beforehand by the user. The first step of the algorithm is to select k random data points as the center of each cluster from the data space D, which might comprise records that are not part of the training set S. Then, the other data points (or records) are assigned to the nearest center. The third step is to compute the gravity center of each cluster. These k gravity centers are the new centers for the clusters. The algorithm then repeats until it reaches stability, which means that none of the data points at S change of cluster from an iteration to another or that the intraclass inertia is now smaller than a certain threshold. Algorithm 5 details the K-Means process:

Algorithm 5. K-Means algorithm.

Input: S the dataset, k the number of clusters to create

Output: Set of k clusters

```
Set the intraclass inertia I_w = ∞
Select k center points c_j ∈ D
Repeat
    For (j ∈ {1, ..., k})
        Set cluster G_j = ∅
    End
    For (i = 1 to |S|)
        Set j* = argmin_{j ∈ {1,...,k}} d(s_i, c_j)
        Set G_j* = G_j* ∪ s_i
    End
    For (j ∈ {1, ..., k})
        Set c_j = gravity center of G_j
    End
    Calculate I_w
Until I_w < threshold
```

To really understand the algorithm, two concepts must be specified—the gravity center and the intraclass inertia. The center of gravity of a dataset X described by p features (attributes) is a synthetic data equal to the average a of each attributes in X:

$$center\ of\ gravity = (a_1, a_2, \dots, a_p) \tag{12}$$

The inertia of a dataset X of $|X|$ records is defined by equation 13:

$$I_X = \sum_{i=1}^{|X|} d^2(x_i, g) \tag{13}$$

where g is the gravity center of X and x_i the ith record of the dataset. Function d^2 represents the Euclidian distance. Finally, the intraclass inertia I_w is given by the following calculation:

$$I_w = \sum_{i=1}^{k} w_i I_i \tag{14}$$

where w_i is the weight of the ith cluster and I_i its inertia. If the data have all the same weight, this weight is calculated by using the number of element members of the cluster G_i and using the formula bellow:

$$w_i = |G_i|/|X| \tag{15}$$

The remainder of the subsection will explain how K-Means works through a visual example. Suppose the dataset is visually represented in a Cartesian plane as shown in Fig. 8(a). In this example, the goal is to find three clusters, so the parameter k is set to three. The Fig. 8(b) shows a possible initialization at the center points of these three clusters. The records of the dataset are assigned to the nearest center.

Then, as explained, the centers of gravity of each cluster are computed from the instances they contain. Data records are reassigned accordingly from their distance to the new centers (see Fig. 9(a)). Finally, the process is repeated until stability is reached. Figure 9(b) shows the final clusters in our example.

The K-Means algorithm is a fast algorithm of linear complexity. In fact, it is considered as one of the fastest clustering algorithms and usually requires a small number of iterations to find the final clusters (Ian H. Witten and Franck 2010). However, there are many drawbacks to the exploitation of this algorithm. First of all, the final clusters are highly dependent on the initial centers that were selected semi-randomly; second, the algorithm converges to local minima. That is, the centers of each cluster move towards a reduction in the distance from their data but there is no guarantee that the global distance will be minimal.

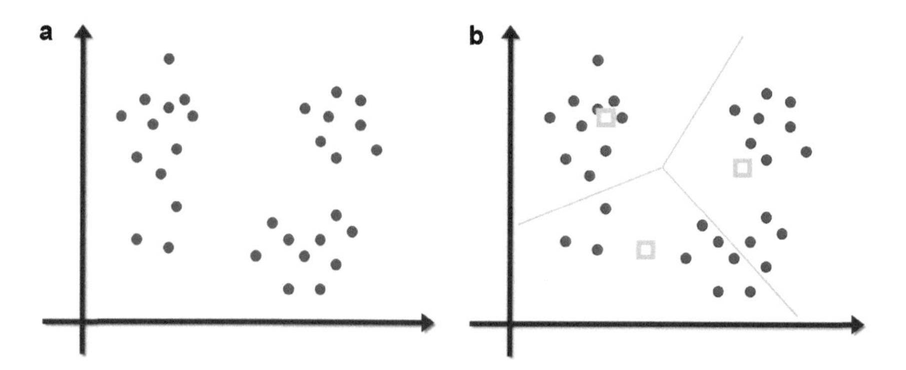

Fig. 8. (a) Dataset before the beginning. (b) Example of initialization with three clusters.

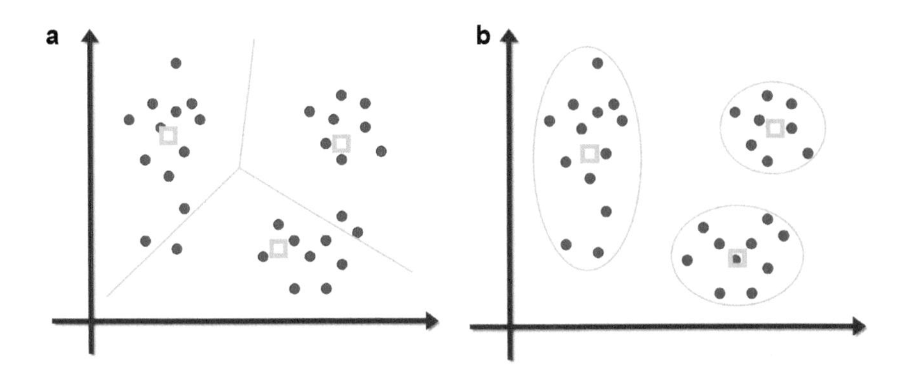

Fig. 9. (a) New clusters after calculation of the new centers. (b) Final clusters.

An improved version named K-Means++ was introduced (Ian H. Witten and Franck 2010) and has to select the center of the first cluster so that it has a uniform probability distribution. Then, the subsequent centers are determined in a manner that their position is proportional to the square of a certain distance value from the first one. That enhancement improves the execution speed and also the precision of the results. However, there is a last problem that remains. To work, the user of K-Means must know the number of clusters beforehand. In the smart home context, it is usually not possible to do so since the number of ADLs that have been realized in the training dataset is not part of our prior knowledge. Therefore, a clustering algorithm that does not require to specify k number of clusters is required.

Density-based Clustering

The first spatial data mining algorithm presented is named Density-Based Spatial Clustering of Applications with Noise or simply DBSCAN (Ester et al. 1996). It is a clustering algorithm that supports noise in the dataset. The goal of this algorithm is to address two of the problems of K-Means based algorithms. First are the weirdly-shaped clusters that cannot be recognized with K-Means; the second is the noise that is necessarily assigned to one of the clusters with K-Means algorithm. Figure 10 shows three sample datasets taken directly from an example of Ester's original paper. A human can easily find the clusters just by looking at each dataset, but K-Means will give poor results on the latest two.

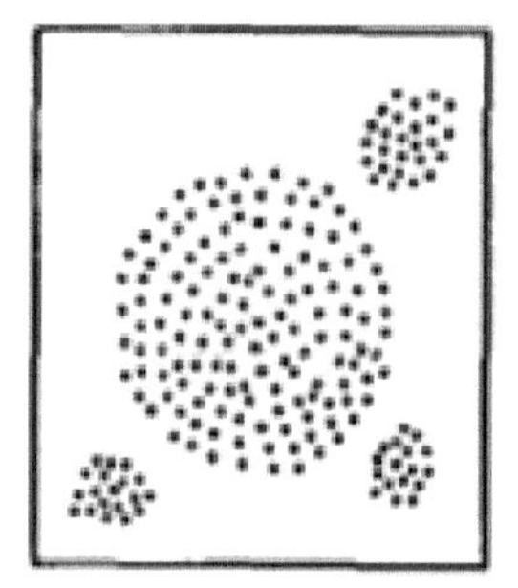

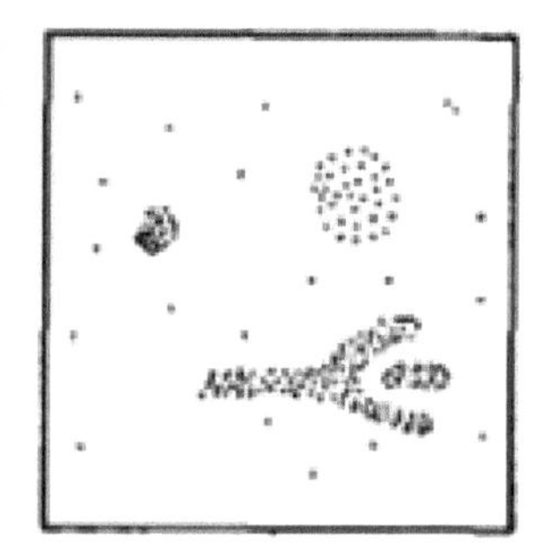

Fig. 10. Three samples dataset from the original paper by Ester.

DBSCAN is based on four important definitions to establish the notion of dense clusters of points. The first definition is $\epsilon - neighborhood$ of a point which comes from mathematical topology:

$$N_\epsilon(p) = \{q \in D | dist(p,q) \leq \epsilon\} \tag{16}$$

This equation describes that q is in the $\epsilon - neighborhood$ of p if the distance between them is smaller than ϵ. An intuitive notion of a dense cluster would be to say that each point has at least *MinPoints* in their $\epsilon - neighborhood$, but it would fail because there are core points and border points in a cluster. The second definition introduced by the team of Ester describes the notion of directly density-reachable point p from a point q:

$$p \in N_\epsilon(q) \ and \ |N_\epsilon(q)| \geq MinPoints \tag{17}$$

That means that p is directly density-reachable from q if it is in its neighborhood and q is a core point (second condition). The relation is symmetric if both points are core type. Third, the point p is density-reachable from a point q if there is a chain of points $P_1, \ldots, P_n, P_1 =\cdot q, P_n = p$ such that P_{i+1} is directly density-reachable from P_i. Finally, a point p is density-connected to a point q if there is a point o such that both p and q are density-reachable from o. Using these four definitions, the authors define a dense cluster as a set of density-connected points. A special set is used to contain the noise; It includes the points that do not belong to any cluster. Figure 11 shows visually the concept density reachability and density-connectivity. Algorithm 6 gives the general idea of the clustering from the concepts presented.

Overall, DBSCAN possesses two important advantages—first, it can be used for applications with noisy data; second, the clusters can be of varied shape: circular, rectangular, elongated, concave, etc. There is also a generalized version (GDBSCAN) (Sander et al. 1998) which allows use of the algorithm with different distances and with two dimensional-shapes. DBSCAN poses some limitations for AR. It is not fast enough for online use. Additionally, it is made for static spatial information rather than changing spatial information such as what is obtained in smart homes. Therefore, it cannot extract the patterns of movement of the various objects in the realization of ADLs.

Algorithm 6. DBSCAN Algorithm.

Input: S the dataset, $mpts$ the minimum number of points, ϵ the neighborhood

Set $clusterID = nextID(NOISE)$
For $(i = 1\ to\ |S|)$
  Set $p = S[i]$
  If $(p.ClID = UNCLASSIFIED)$ Then
    If $(ExpendCluster(S, p, clusterID, \epsilon, mpts))$ then
      Set $clusterID = nextID(clusterID)$
    End
  End
End

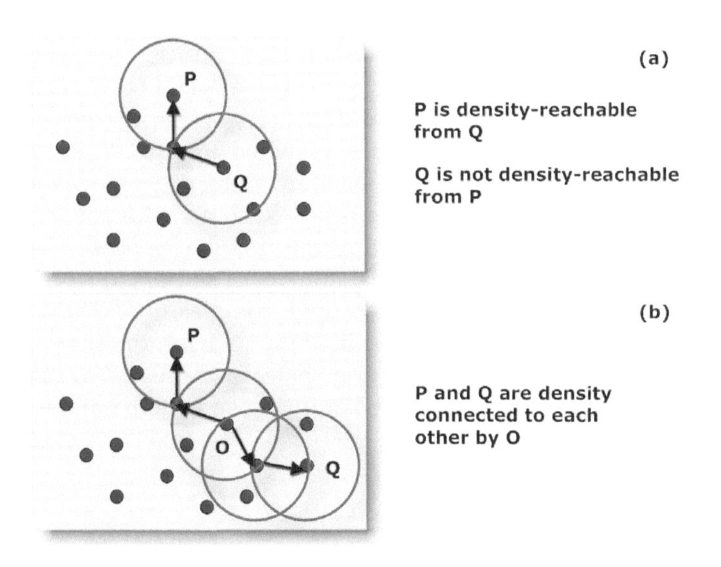

Fig. 11. (a) Density-reachability. (b) Density-connectivity.

Moving clustering

The last algorithm we present has a very different approach to clustering. The algorithm presented in (Liu et al. 2010) aims to develop a mobility-based clustering for the monitoring of vehicles' crowdedness in metropolis. Their idea is to use solely the current speed of vehicles since a high mobility means low crowdedness. The main challenge of their approach is not one of clustering; in fact, deals with contextual information (e.g. red light, etc.) and the imprecision of GPS data. Otherwise, most of their work is based on statistical methods. There are many advantages to mobility-based clustering. First, it is little sensitive to the size of the sample; second, it does not require precise position and support errors in positioning; and finally, it naturally incorporates the mobility of different objects such as vehicles. Even though the model is not general enough to be directly applied to our problem, we found the idea innovating and it led us in the quest for a new spatial data mining method. Still, mobility-based clustering is new and much work remains to be done to obtain interesting accuracy.

The bottom line

Clustering seems to be a very good opportunity for AR, but only few approaches have successfully exploited it (Wyatt et al. 2005). Moreover, every time it is with a small number of low granularity activities. For example, Palme et al. (2010) used a completely unsupervised method that extracted the most relevant object to represent an ADL (key object). It is limited by the uniqueness required of the key object. In general, however, there are many reasons that explain why very few approaches exist. First of all, the complexity of information gathered from multiple sensors in smart homes limit the ability of a standard clustering algorithm to sieve correctly the data. In fact, an algorithm, such as K-Means, is not able to distinguish noise from interesting information. Second, most of the clustering algorithms need the initial number of clusters to work correctly and those, that do not, are very slow (high complexity). Finally, standard clustering algorithms do not fully exploit the ADL information embedded in the dataset; for example, they ignore many fundamental spatial aspects such as the topological relationships or the movement of entities.

4. Conclusion

Smart homes present a good opportunity to address challenges that our society will face in the future. However, there are numerous challenges that researchers and engineers need to address in order to implement the dream; in particular, recognizing the ongoing activities of daily living has proven a difficult task to do. Classical approaches exploited mathematical formalisms, such as the first-order logic or a probabilistic model like the Bayesian network. One limitation of the models is the requirement for a human expert to build a plans' library. That plans' library is usually assumed to be complete and correct which strongly limits its real-world application. Data mining can help in solving this issue. The three main families of data-mining approaches were presented in this chapter. Each of them possess their strengths and their weaknesses, but the main thing to remember is that it is often better to adapt the solution to the particular problem/context. In fact, *ad hoc* solutions often work better in smart homes than theoretical frameworks. However, this is a temporary solution, since smart homes need to scale easily in the future. Further effort is needed from the research community regarding data mining approaches. In particular, it is important to explore incremental data mining and focus on the generalization of solutions to multiple environments and multiple users. Finally, remember that data mining is not a silver bullet; it will not magically solve all problems in smart homes.

Keywords: Smart home, data mining, decision tree, clustering, association rules

References

Agrawal, Rakesh and Ramakrishnan Srikant. 1994. Fast algorithms for mining association rules in large databases. pp. 487–99. *In*: Proceedings of the 20th International Conference on Very Large Data Bases. Morgan Kaufmann Publishers Inc.

Albrecht, David W., Ingrid Zukerman and N.E. Nicholson. 1998. Bayesian models for keyhole plan recognition in an adventure game. User Modeling and User-adapted Interaction. 8(1-2): 5–47. doi: 10.1023/a:1008238218679.

Allen, James F. and Patrick J. Hayes. 1985. A common-sense theory of time. pp. 528–31. *In*: Proceedings of the 9th International Joint Conference On Artificial Intelligence - Volume 1, Los Angeles, California: Morgan Kaufmann Publishers Inc.

Bao, Ling and Stephen S. Intille. 2004. Activity recognition from user-annotated acceleration data. pp. 117. *In*: Pervasive Computing. Springer.

Barlow, Horace B. 1989. Unsupervised learning. Neural Computation. 1(3): 295–311.

Baum, C. and D.F. Edwards. 1993. Cognitive Performance in Senile Dementia of the Alzheimer's Type: The Kitchen Task Assessment. Vol. 47.

Bouchard, B., S. Giroux and A. Bouzouane. 2007. A keyhole plan recognition model for Alzheimer's patients: First results. Journal of Applied Artificial Intelligence. 21(7): 623–58. doi: 10.1080/08839510701492579.

Bouchard, Bruno, Patrice Roy, Abdenour Bouzouane, Sylvain Giroux and A. Mihailidis. 2008. An activity recognition model for Alzheimer's patients: extension of the COACH task guidance system. pp. 811–2. *In*: Malik Ghallab and D. Constantine (Eds.). ECAI. Spyropoulos, Nikos Fakotakis and Nikolaos M. Avouris, IOS Press.

Bouchard, Kevin, Bruno Bouchard and Abdenour Bouzouane. 2013. Discovery of topological relations for spatial activity recognition. Paper presented at the IEEE Symposium Series on Computational Intelligence, Singapore.
Carbonell, Jaime G., Ryszard S. Michalski and Tom M. Mitchell. 1983. An overview of machine learning. pp. 3–23. *In*: Ryszard S. Michalski, Jaime G. Carbonell and Tom M. Mitchell (Eds.). Machine Learning. Springer Berlin Heidelberg.
Chaudhuri, Surajit, Umeshwar Dayal and Vivek Narasayya. 2011. An overview of business intelligence technology. Commun. ACM. 54(8): 88–98. doi: 10.1145/1978542.1978562.
Chen, Liming, Chris D. Nugent and Hui Wang. 2012. A knowledge-driven approach to activity recognition in smart homes. IEEE Trans. on Knowl. and Data Eng. 24(6): 961–74. doi: 10.1109/tkde.2011.51.
Cohen, Philip R. 1984. The pragmatics of referring and the modality of communication. Computational Linguistics. 10(2): 97–146.
Egenhofer, M.J. and R.D. Franzosa. 1991. Point-set topological spatial relations. International Journal of Geographical Information Systems. 5(2): 161–74.
Ester, Martin, Hans-Peter Kriegel, Jörg Sander and Xiaowei Xu. 1996. A Density-based Algorithm for Discovering Clusters in Large Spatial Databases with Noise.
Gaddam, Shekhar R., Vir V. Phoha and Kiran S. Balagani. 2007. K-means+ id3: A novel method for supervised anomaly detection by cascading k-means clustering and id3 decision tree learning methods. Knowledge and Data Engineering, IEEE Transactions on. 19(3): 345–54.
Gopalratnam, Karthik and Diane J. Cook. 2003. Active LeZi: An incremental parsing algorithm for sequential prediction. pp. 38–42. *In*: Ingrid Russell and Susan M. Haller (Eds.). FLAIRS Conference. AAAI Press.
Gu, Tao, Shaxun Chen, Xianping Tao and Jian Lu. 2010. An unsupervised approach to activity recognition and segmentation based on object-use fingerprints. Data Knowl. Eng. 69(6): 533–44. doi: 10.1016/j.datak.2010.01.004.
Ian H. Witten and Eibe Franck. 2010. Data Mining: Practical Machine Learning Tools and Techniques. Elsevier editor.
Jakkula, V. and D.J. Cook. 2007. Learning temporal relations in smart home data. Paper presented at the Proceedings of the Second International Conference on Technology and Aging, Canada.
Jakkula, V.R. and D.J. Cook. 2008. Enhancing smart home algorithms using temporal relations. pp. 3–10. *In*: Mihailidis, A., J. Boger, H. Kautz and L. Normie (Eds.). Technology and Aging. Amsterdam: IOS Press.
Kasteren, Tim van, Athanasios Noulas, Gwenn Englebienne and Ben Krose. 2008. Accurate activity recognition in a home setting. pp. 1–9. *In*: Proceedings of the 10th international conference on Ubiquitous computing. Seoul, Korea: ACM.
Katz, Sidney, Amasa B. Ford, Roland W. Moskowitz, Beverly A. Jackson and Marjorie W. Jaffe. 1963. Studies of Illness in the Aged. JAMA: The Journal of the American Medical Association. 185(12): 914–9. doi: 10.1001/jama.1963.03060120024016.
Kautz, Henry A. 1991. A formal theory of plan recognition and its implementation. pp. 69–124. *In*: Reasoning about plans. Morgan Kaufmann Publishers Inc.
Komninos, Andreas and Sofia Stamou. 2006. HealthPal: an intelligent personal medical assistant for supporting the self-monitoring of healthcare in the ageing society. Proceedings of the UbiHealth.
Liu, Siyuan, Yunhuai Liu, Lionel M. Ni, Jianping Fan and Minglu Li. 2010. Towards mobility-based clustering. pp. 919–28. *In*: Proceedings of the 16th ACM SIGKDD International Conference on Knowledge Discovery and Data Mining. Washington, DC, USA: ACM.
Meir, Ron and Gunnar Rätsch. 2003. An introduction to boosting and leveraging. pp. 118–83. *In*: Advanced Lectures on Machine Learning. Springer.

Nguyen, Nam, Hung Bui, Svetha Venkatesh and Geoff West. 2003. Recognising and Monitoring High-Level Behaviours in Complex Spatial Environments. Paper presented at the In IEEE International Conference on Computer Vision and Pattern Recognition (CVPR).
Nugent, Chris, Maurice Mulvenna, Ferial Moelaert, Birgitta Bergvall-Kareborn, Franka Meiland, David Craig, Richard Davies et al. 2007. Home based assistive technologies for people with mild dementia. pp. 63–9. *In*: Proceedings of the 5th International Conference on Smart Homes and Health Telematics. Nara, Japan: Springer-Verlag.
Palmes, Paulito, Hung Keng Pung, Tao Gu, Wenwei Xue and Shaxun Chen. 2010. Object relevance weight pattern mining for activity recognition and segmentation. Pervasive Mob. Comput. 6(1): 43–57. doi: 10.1016/j.pmcj.2009.10.004.
Quinlan, J. Ross. 1996. Bagging, boosting, and C4. 5. Paper presented at the AAAI/IAAI, Vol. 1.
Quinlan, J.R. and J. Ghosh. 2006. Top 10 Algorithms in Data Mining. Paper presented at the International Conference on Data Mining (ICDM'06), Hong Kong, December 18–22.
Ravi, Nishkam, Nikhil Dandekar, Preetham Mysore and Michael L. Littman. 2005. Activity recognition from accelerometer data. Paper presented at the Proceedings of the National Conference on Artificial Intelligence.
Ricquebourg, V., D. Durand, L. Delahoche, D. Menga, B. Marhic and C. Loge. 2006. The Smart Home Concept: Our Immediate Future. Paper presented at the 1ST IEEE International Conference on E-Learning in Industrial Electronics.
Robles, Rosslin John and Tai-hoon Kim. 2010. Applications, systems and methods in smart home technology: a review. Int. Journal of Advanced Science and Technology. 15: 37–48.
Selim, S.Z. and M.A. Ismail. 1984. K-means type algorithms: a generalized convergence theorem and characterization of local optimality. IEEE Transactions on Pattern Analysis and Machine Intelligence. 6: 81–7.
Sander, Jörg, Martin Ester, Hans-Peter Kriegel and Xiaowei Xu. 1998. Density-based clustering in spatial databases: The algorithm gdbscan and its applications. Data Mining and Knowledge Discovery. 2(2): 169–94.
Schwartz, Myrna F., Mary Segal, Tracy Veramonti, Mary Ferraro and Laurel J. Buxbaum. 2002. The Naturalistic Action Test: A standardised assessment for everyday action impairment. Vol. 12. Hove, ROYAUME-UNI: Psychology Press.
Srikant, Ramakrishnan and Rakesh Agrawal. 1996. Mining Sequential Patterns: Generalizations and Performance Improvements. pp. 3–17. *In*: Proceedings of the 5th International Conference on Extending Database Technology: Advances in Database Technology. Springer-Verlag.
Stankovski, V. 2006. Application of decision trees to smart homes. *In*: Designing Smart Homes. Springer.
Todorovski, Ljupčo and Sašo Džeroski. 2003. Combining classifiers with meta decision trees. Machine Learning. 50(3): 223–49.
Turaga, P., R. Chellappa, V.S. Subrahmanian and O. Udrea. 2008. Machine Recognition of Human Activities: A Survey. IEEE Transactions on Circuits and Systems for Video Technology. (11): 16.
Weiss, Gerhard. 2000. Multiagent Systems: MIT Press.
Wyatt, Danny, Matthai Philipose and Tanzeem Choudhury. 2005. Unsupervised activity recognition using automatically mined common sense. pp. 21–7. *In*: Proceedings of the 20th National Conference on Artificial Intelligence - Volume 1. Pittsburgh, Pennsylvania: AAAI Press.

7

Preliminary Evaluation of a Digital Diary for Elder People in Nursing Homes

Laetitia Courbet, Agathe Morin, Jérémy Bauchet* and *Vincent Rialle*

1. Introduction

1.1 Purpose

Acceptance is a major issue to overcome usage process in assistive technology. A great amount of studies demonstrated a low clinical value for digital assistive devices despite the fact that they have had already been robustly evaluated as efficient and useful. We present usability and acceptability study of a web based digital diary for online use by old people with Mild Cognitive Impairments (MCI), and designed to improve their quality of life (QoL). The study was conducted within the European project MyGuardian (2012–2015), an Information System (IS) research project within the 'Ambient Assisted Living' (AAL) European program. It aimed to reduce the impact cognitive impairments have on MCI peoples' daily life, with a focus on outdoor life and mobility, and also considering their caregivers.

Université Grenoble Alpes, EA n°7407 AGEIS Laboratory, Grenoble, France.
* Corresponding author: laetitia.courbet@agim.eu

The interest of reducing this impact was supported by recent studies. For instance, Wettstein et al. (2014) demonstrated the moderating effect of cognitive status on the relationship between out-of-home behaviour, oneself sensation of environmental mastery. On the other hand, as an expression of the resistances in connection with the "need" to be helped, the acceptance level of digital devices appears to be critical (Korall et al. 2015). This acceptance is a dynamic, fluctuating and multifactorial process that requires the acknowledgement of the acquired vulnerability (Peeks et al. 2014) as well as a permissive social influence (Korall et al. 2015). Bypassing the cared person's resistance toward the assistance seems to rely on key concepts like collaboration, and interdependency (Allen and Willes 2014; Fine and Glendinning 2005). The users' comprehension of the device is also central (Laurent et al. 2014). Besides, a device appropriation, as part of the acceptance process, directly depends on the facilitation or constraining effect of technology on individuals' action (DeSanctis and Poole 1994; Orlikowski and Robey 1991). How the technology will be accepted is therefore difficult to anticipate. As a consequence, the assessment of assistive technology acceptance requires a methodology to control all the various and dynamic acceptance factors without constraining the participants' use and judgment of the device.

The MyGuardian project was twofold: first, alarm handling, mobility, safety and security of residents of a nursing home; second, use of social networking mechanisms to enable easy coordination. The study presented thereafter is related to the latter. It focuses on the preliminary evaluation of the above-mentioned MyGuardian digital diary (referred to as 'MG') for online use by old people living in nursing homes. The web interface of this diary features the standard functionalities of classical digital diaries, along with mechanisms to coordinate the assistance provided by the care network, with message services and an edition tool for the safe zones and related alarms.

1.2 Objectives and Hypotheses

Objectives:

- Validate the functional definition of the technological means that permit the removal of the brakes that oppose the minimal mobility for a satisfying social life.
- Identify and eliminate the main brakes that oppose MG usability, from ergonomics and technological perspectives.
- Identify the level of acceptance of MG.

Main hypothesis:

There is a correlation between a senior's mobility and the confidence he has toward himself and toward how he interacts with his environment.

Secondary hypotheses:

- Bypassing the cared person's resistance toward the assistance rely on key concepts like collaboration and interdependency (Allen and Willes 2014; Fine and Glendinning 2005).
- The acceptance of assistance by a person with an acquired vulnerability is optimized by her close relationships presence, participation and support (Chan et al. 2000; Taverner-Smith and DeVet, 2006; Honkanen et al. 2007; O'Halloran et al. 2005 as cited in Korall et al. 2015).
- The acceptance is optimized by the persons' comprehensions of the device (Laurent et al. 2014).
- The optimization of the acceptance factors dynamic results in the optimization of the quality, and the quantity of the out comings approvals, rejections and (un)ease of use markers and therefore on a more reliable and rich evaluation.

2. Method

2.1 Population

Twelve participants were included in the study (Table 1): 9 women (84, 8 ± 7 years old) and 3 men (79 ± 7 years old). Five of them were living in a retirement home, all women; 7 of them were living in community dwelling (3 men, 4 women, 4 of them were couples). Data are gathered from elderly people with the following inclusion and ethics criteria:

- Over 65 years old.
- Informed consent to participate to the study.

Recruitment was done thanks to the retirement home staff and a nursing office. All participants were from Grenoble area, France.

Mean duration of the encounters was 31 minutes (± 12 min).

Table 1. Population characteristics.

	Gender	Age	Family situation	Place	Technological equipment	Interview duration (mn)
1	man	88	single	at home	yes: computer	49
2	woman	92	single	at home	no	12
3	man	80	couple	at home	yes: computer	41
4	woman	79	couple	at home	yes: computer	28
5	woman	93	single	at home	no	38
6	woman	91	single	retirement home	no	35
7	woman	88	single	retirement home	no	30
8	woman	81	single	retirement home	yes: computer	5
9	woman	67	couple	at home	no	34
10	man	69	couple	at home	no	30
11	woman	85	single	retirement home	no	35
12	woman	88	single	retirement home	no	34

2.2 Method

Individual encounters were organized in 4 steps as follows:

Step 1: presentation of the study and collect of the informed consent. This step aims to introduce the participants with the study and make them feel comfortable with the purpose of the encounter.

Step 2: evaluation of the participant's appetence for new technologies.

Step 3: testing the web application usability. Each participant was asked to complete a predefined list of tasks including the functionalities provided by the agenda. Level of independence when completing the listed tasks was evaluated thanks to a 6 items scale. Tools developed by occupational therapist, and used for the independent living skills assessment inspired this scale (Dutil et al. 1996). The Dutil's scale is usually used to develop models for different cognitive tasks. Its measure is based on the level of assistance required to successfully complete the target task. It is evaluated according to six criteria on the clinical dementia rating (CDR): CDR 0 (independent), CDR 0.5 (success with confirmation), CDR 1 (success with incitement), CDR 2 (success with guidance), CDR 3 (success with guidance and slowness of execution), CDR 4 (failure). For each criterion, the level of support provided by the test administrator is scored from 0 to 3 (0 for a subject who performs without any help, 1 for a subject who requires a verbal cue, 2 for a subject who requires physical assistance, and 3 for a subject totally unable to perform that section of the test). All subjects who completed were able to respond to verbal or physical assistance. There was no time limitation to complete the different tasks, and the participant was allowed to ask for assistance to the experimenter.

Nine tasks were proposed to the person:

* *First task*: find the starting up button the computer and switch it on.
* *Second task*: (passive consultation) on the homepage, observe attentively the general menu (icons, images, text) and explain the purpose of the various elements that you see?
* *Third task*: go to the diary. Observe attentively the page and explain the purpose of the various elements that you see (passive consultation).
* *Fourth task*: Once on the diary, create an appointment at the hairdresser on January 10th 2015, at 4 pm. Go to the right date and select the right hour. Put a title to your appointment.
* *Fifth task*: now can you modify the appointment? Go to your appointment and change the date or the hour.
* *Sixth task*: now can you delete the appointment?
* *Seventh task*: please create a new task: water the plants Please save your task.
* *Eighth task*: please read your messages and answer in the course of the discussions.
* *Last task*: disconnect from the diary and put out the computer/the tablet.

Step 4: evaluation of the participant's feeling about the web application. The first step consisted in a self-administered questionnaire composed with the following eight questions: 1. In your opinion the tasks are? (Multiple-choice question: [Easy, Rather easy, Difficult, Very difficult, Impossible to be done]). 2. Have you been comfortable with the exercises? 3. Would you have needed help? 4. The laptop manipulation is? (Multiple-choice question: idem q. 1). 5. Do you feel this digital diary can be useful for you? 6. Would you be willing to take ownership of a laptop or a touchpad? 7. Do you find that the digital diary is rather clear and understandable? 8. Are the given instructions clear enough to you?

Encounters were done at the participant's home. The experimenter came with a laptop or the participant own computer was used to test the web application. A connection to the Internet was mandatory.

At the end of the encounter, participants were asked to use the web application during 3 days on a tablet that was provided by the experimenter. A semi-directed interview was done at the end of the 3 days test period.

2.3 Inclusion Criteria

The inclusion criteria are summarized as follows:

- Elderly with a diagnosis of MCI (the underlying etiologic diagnosis, if possible posed, is not a criterion).
- Senior citizen with the level of independent mobility (travel alone and on foot), at least once a week (short ballads, small purchases of daily...).
- Adult persons residing at home, members or beneficiaries of a security regime Health.
- Motivation for the study, interest in the issues addressed, consent to participation of the entire dyad.

The non-inclusion criteria are summarized as follows:

- Significant disturbances of walking and balance.
- Behavioural disorders incompatible with the type of study (depression or important apathy).
- Reluctance to use new technologies, refusal of consent of one of the members or of the whole of the dyad.

Comments on the inclusion and non-inclusion criteria:

- Anosognosia is not a criterion of non-inclusion.
- A complaint specifically relating to the spatio-temporal orientation is not indispensable.
- Wandering is not a criterion of inclusion.
- The age of the senior was not an inclusion criterion.

2.4 Ethical Issues

This study was not an "interventional" one according to the French bioethical laws. The intervention was limited to verbal exchanges addressing the real usage sessions of MG. This research did not impact the physical and moral integrity of the participants. An administrative procedure was lead with the French ethical authority CPP SUD-EST II (Comité de Protection des Personnes SUD-EST II).

3. Results

3.1 Computer Usability Outcomes

8 upon 12 participants never used a computer before the study. Nevertheless, their participation was important for comparison purposes, and also to enlarge the sample size.

Manipulating the computer was a challenge for most of the participants. To avoid the difficulties induced by using the computer's pointing device, we asked the participants having such difficulties to show directly on the screen, with their fingers, where to point and click. Unfortunately, for those persons, ending the tests was not possible because of the level of concentration needed, and the resulting tiredness.

3.2 Digital Diary Usability Outcomes

2 tasks requested were passive tasks, meaning that the participant was asked to describe what he saw on the screen, and what was, according to him, the purpose of the web application. These tasks were not evaluated.

Two participants said that the purpose of the web application was explicit, that the "task" part was planned to note what they had to do, and that the "diary" part aimed to manage a meeting. Some of them saw the interface as a "journal". For 2 participants, the application had no utility, and could be considered as a pastime. The reason was that it took more time for them to complete the information on the screen than to write them on a paper agenda, or on a calendar (regarding the appointments), or on a notebook (regarding things to do).

Most of participants did not master the time chronology concept used by the application, and they needed a lot of time to complete the tasks. They randomly clicked on the web interface components during the completion, and did not really understand the meaning of the elements. Most of them did not read what was written on the screen, they tried to find the correct way to complete the task as quickly as possible, and as consequence they selected several items at the same time without focusing on the task the experimenter asked them to complete. They were therefore set to fail, and this matter of fact induced annoyance and frustration. Their conclusion was that they did not have the basic knowledge to do the tasks correctly, even if several participants expressed that completing the tasks was easy.

Most of the participants said that for a first use, the application implied several complicated manipulations. Some of them agreed to use the application another time with some assistance.

No participants completed independently the **first task** (switch on the computer) and the **last task** (disconnect from the application and switch off the computer). Table 2 shows a general overview of the results of the 12 participants:

The italic numbers in the Table 2 represent remarkable results: some are high (5 and 6) when compared to the intermediate numbers (ranging from 1 to 4), or are relatively very low (0). The **second** and the **third tasks** are empty because they were passive tasks that didn't need to be evaluated. For the **last task**, nobody succeeded. Moreover, the few people who succeeded to go to the end of the **ninth task** were peculiarly tired.

3.3 Main Difficulties Met

- Including participants was difficult as most considered that computers were not for them; sometimes the opposition came from the family who expressed that such a tool was not suitable for their relative.
- Ability to manipulate the pointing device.
- Some of the participants stopped during the tests, saying that they won't be able to complete the tasks, and that it had no interest.
- Impact of eyesight impairments on the tasks completion, conducting one participant not to complete the whole study.
- Impact of articular impairments (degenerative joint disease) on the ability to use the computer, conducting one participant not to complete the whole study.
- Impact of cognitive impairments on the ability to understand the tasks that had to be completed.

Table 2. Tasks requested from participants.

	0 = Independent	0.5 = Success with confirmation	1 = Success with incitement	2 = Success with guide	3 = Success with guide and slowness of execution	4 = Failure
<u>First task</u>:						
* Find the button of starting up of the computer	*5/12*	1/12	1/12	1/12	1/12	3/12
* Switch it on						
<u>Second task</u>:						
* Passive consultation of the HOME page						
<u>Third task</u>:						
* Passive consultation of the Diary interface						
<u>Fourth task</u>:						
* Add an appointment	1/12	2/12	2/12	2/12	*0/12*	*5/12*
* Find the good date	2/12	1/12	2/12	1/12	3/12	3/12
* Select the good hour	2/12	*0/12*	2/12	2/12	3/12	3/12
* Put a title in your meeting	1/12	1/12	1/12	3/12	3/12	3/12
* Protect the meeting	1/12	4/12	*0/12*	*0/12*	2/12	*5/12*
<u>Fifth task</u>:						
* Modify your meeting: * Change the date or another thing	1/12	1/12	*0/12*	4/12	2/12	4/12

* Protect the modification	2/12	1/12	0/12	3/12	1/12	5/12
Sixth task:						
* Delete this appointment	2/12	0/12	2/12	1/12	3/12	4/12
Seventh task:						
Create a new task * Spray plants	1/12	0/12	4/12	2/12	2/12	3/12
* Give the title of "spray plants" in your task	1/12	0/12	4/12	1/12	3/12	3/12
* Protect the task	2/12	3/12	0/12	1/12	2/12	4/12
Eighth task:						
* Write a message	1/12	1/12	1/12	0/12	5/12	4/12
Ninth task:						
* Repeat this task in the tablet	1/12	1/12	0/12	2/12	2/12	6/12
Last task:						
* Disconnect from the application and switch off the computer						

- Regarding the digital diary:
  - The first name is sometimes written with only the two first letters of the name (for appointments that are accompanied). Participants did not understand that these letters stand for the first name of the person that should accompany them.
  - When asking to write a message, participants selected the "send message" icon instead of the text field.
  - The "home" button was not explicit. Going back to the home page was therefore inefficient.
  - As the "title" text field provided a default sentence, participants entered the title in the description text field.
  - The "save" button was not explicit enough.

3.4 User's Feeling Toward the Protocol and the Technology: The Outcomes

The participant answers to the users' feelings questionnaire. Given the current little number of participants who filled that questionnaire in, these results aren't significant and are to be processed as qualitative results.

1. In your opinion, the tasks are:
   For the majority (5 people) tasks are « rather easy », three answered « easy », two answered « very difficult », and the two other persons answered « difficult » and « impossible to done ».
2. Have you been comfortable with the exercises?
   Nine persons answered "yes" and three answered "no".
3. You would have needed help?
   Eight persons answered "yes" and four answered "no".
4. The laptop manipulation is:
   For six persons, it is easy. (50%)
   For three persons, it is rather easy. (25%)
   For three persons, it is difficult. (25%)
5. This application can be useful for you?
   Eight people answered "yes" and four answered "no".
6. Would you be willing to take ownership of a laptop or a tablet?
   Six persons answered "yes" and six other persons answered "no". (50%/50%)

Additional comments:

Investigator: "And if I install the digital diary on your own device?"

Participant: "No as I don't need it, I have all my appointments in my mind. I am enough autonomous to remember my appointments"

7. This application is rather clear and understandable?
   For seven persons, it is easy.
   For one person, it is rather easy.
   For four persons, it is difficult.

Additional comments:

Participant: "It is clear enough but this application is not suitable for me, as there are main things in the application that I don't catch right from the beginning"

8. Does the given instruction seemed clear enough to you?
   For ten persons, it is easy.

On two persons, one answered that it is rather easy, and the other person answered that it is difficult.

Considering the additional comment, it appears that the participants don't give an answer to the instructions formulations but rather on their feasibility on the application. The researcher needs to stress the participants' answer consistently with this outcome.

3.5 Results Regarding the use of the Web Application on a Tablet during 3 Days

On the 12 participants, 3 accepted to use the web application during 3 days on a tablet provided by the experimenter. One person was hospitalized and could not complete the 3 days test. Another one had cognitive impairments that were too important to have significant results. As a consequence, only one person was interviewed at the end of the 3 days test. This use case is presented below.

1. Have you used MyGuardian application on the tablet during these 3 days?
   I did.

2. How many times?
   Two times. The first one alone, the second one with the assistance of my daughter.
3. How longer did you use it?
   The first time only 10 minutes, just to see if I could turn the tablet on alone, but I could not as the passwords were required to be connected. The second time 30 or 40 minutes with my daughter and my wife.
4. What have you done with the tablet?
   We tried to complete the same exercises as during the first test.
5. Was it easy for you to use the application?
   It wasn't. I think it's not very ergonomic, icons and images are small, and also is the keyboard that appears suddenly when you want to write something and that takes half of the screen. Moreover, the keys are really too small and too sensitive.
6. Do you think this application could be useful for you?
   I don't think so, as I never used a computer before I think I'll have difficulties. But regarding the concept I think it's useful, as less papers will lie around and we can carry it easily and all information will be gathered on only one support.

   I think it could more useful for people that already master computers; I don't and learn something new today will be difficult for me.
7. What are the brakes for using the tablet and the application?
   - *It's something new for me.*
   - *I prefer to speak directly with someone face to face and also I'm afraid making a mistake when using it that is to send a message to someone who was not the message recipient.*
8. Would you have used it if some support were provided?
   I used it with my daughter so I would, but certainly more as a pastime than for a serious usage.
9. What was displeasing regarding the interface?
   Nothing. It's just too sophisticated for someone like me.
10. What should be done on the tablet so that it will be easy to use?
    What I said you before. I like the speech synthesis, for someone like me that has difficulties to use the keyboard I think it's useful.

4. Discussion

Whether participants were familiar with technology or not, no major outcome came out from the passive consultations: tasks 2 and 3 in the step 2 of the protocol. These passive consultation tasks revealed problems regarding the global aspects of the interface (perceived visibility, simplicity, good looking, primary ergonomic elements…). As no major outcomes came out during this part of the protocol, it seems that the interface was well accepted at first glance.

Nevertheless, the usability tests allowed detecting some ergonomic problems:

1) The participants experienced difficulties to orientate themselves in the Home interface, and to memorize which bloc corresponded to the various parts of the interface (diary, tasks, and messages).
2) In the French version, some words were not sufficiently adapted to the elder's comprehension, and needed to be replaced by more familiar ones.
3) In the appointment creation, the old persons did not know what to do with the "Description" box: they did not spontaneously understand its purpose.
4) There was an important readability issue: The font size was too small, as well as some signs such as "+" standing for an adding, and "←" standing for "back to the home page".

Consistently with these primary outcomes, a few elements of the web application interface would gain to be adjusted. For instance:

- On the 'HOME' interface, associating colors to categories (diary, tasks, messages) might ease the comprehension; contacts: in the French version of the interface, the word "contact" should be replaced by "répertoire";
- On the diary interface: Appointment creation: in the French version of the interface replace "Sauvegarder" by "Confirmer", more known and used by elders. The elders don't know what to put in the "description" box. We therefore propose two potential solutions which would need to be further tested: (1) suppress this box, as the elders don't spontaneously

see the purpose of it, they are questioning at the same time the utility of the box; (2) replace "description" by something more explicit, such as "Complementary details about the appointment", "Détails complémentaires concernant ce RDV" in French.

We proposed the implementation of the 2nd solution so that it could be tested in the handovers to come: add the possibility for the elders to set a reminder to the appointment (a reminder 2 days, 1 day, 1 hour, 15 minutes before it).

Proposals regarding all the interfaces: Increase the font size as well as the "+" sign standing for an adding; modify the sign standing for coming back to the 'Home page'. It was too small, and was not intuitive for elders. We proposed to keep the sign (←), and to make it larger, and to add a full sentence to it, such as "Go back to the HOME screen" ("Retourner à l'écran d'accueil" in French).

Protocol update considering these first handovers: (1) the touchpad needs to be deactivated during the handovers. Only the mouse is to be used. (2) Some sub tasks that will be suppressed from the protocol: **Fifth task:** Invite a person of your choice in your appointment; **Sixth task:** Delete this appointment; **Seventh task:** Archive the task, and invite someone to participate; **Eighth task:** delete your message.

It must be stressed that individual factors (age, education level, and computer experience) were found to affect task performances. A variety of assistive technologies have been conceived and developed to support independent living and quality of life of older adults with MCI or Alzheimer's disease. Our results are comparable to the ones obtained by M. Pino et al. (2012) regarding the assessment of a graphical user interface (GUI) for a social assistive robot for older adults with cognitive impairment.

This article deals with socially assistive robotics, it examines eleven old people with MCI and 11 old with normal cognition were recruited for this study; participants were asked to complete a series of tasks using the main menu of the GUI, and navigate through its different applications. Performance and satisfaction measures were collected (e.g., time to complete each task, number of errors due to manipulation, number of help requests). Tests were conducted individually. Some particular aspects of the interfaces (icons, navigation system) had to be modified to make the application usable by the largest number of patients suffering from cognitive deficits.

5. Conclusion

The MyGuardian digital diary can be useful for people having skills in computer, or who want to learn how to use a computer, provided some amelioration be made. Its main brake is that people have to turn on the computer, and to open the application each day. It means that we have to anticipate the way people will go from the paper agenda to the digital one, the way to integrate it in their daily life, and the human support that is needed. It seems that the application is not adapted to the current generation of senior, but it could be useful to support people having severe memory impairments in their daily life. Furthermore, the application needs an Internet connection, and not all seniors have it today, or they don't know how to use it.

Tests were conducted with people whose average age is 83, who don't want to have to learn new skills, and for which computer is not at the center of their interests. Some of them don't want to learn how to use computer, as they don't understand how it works, or they are afraid of going to fail. As a result, they sometimes expressed some aggressiveness, and irritation probably in order not to show their weakness. The fear of being in a situation of failure explains partially this behaviour. The results were globally satisfactory. No major problems to perform the diverse tasks were identified. However, we noticed that the participants took a lot of time to complete the tasks: execution was slow. As a result, a quite long training period must be planned for old people to use the web application.

Acknowledgement

The MyGuardian project had received funding from the European Union's Ambient Assisted Living program. We thank the French National Agency of Research (ANR) who managed the AAL funding in France. The authors gratefully acknowledge the professionals of retirement homes: they participated to the participant recruitment. Above all, we thank the elders who accepted to be recruited for the study.

Keywords: nursing home, elders, web application, e-health, gerontechnology, frailty

References

Allen, R.E. and J.L. Wiles. 2014. Receiving support when older: what makes it ok? The Gerontologist. 54(4): 670–82.

Chan, D.K., G. Hillier, M. Coore et al. 2000. Effectiveness and acceptability of a newly designed hip protector: A pilot study. Arch. Gerontol. Geriatr. 20: 25–24.

DeSanctis, G. and M.S. Poole. 1994. Capturing the complexity in advanced technology use: adaptive structuration theory. Organization Science. 5: 2, 121–147.

Fine, M. and C. Glendinning. 2005. Dependence, independence or inter-dependence? Revisiting the concepts of care and dependency. Ageing & Society. 25(04): 601–621.

Honkanen, L.A., N. Monaghan, M.C. Reid et al. 2007. Can hip protector use in the nursing home be predicted? J. Am. Geriatr. Soc. 55: 350–356.

Korall, A.M.B., F. Feldman, V.J. Scott, M. Wasdell, R. Gillan, D. Ross et al. 2015. Facilitators of and barriers to hip protector acceptance and adherence in long-term care facilities: A systematic review. Journal of the American Medical Directors Association. 16(3): 185–193.

Laurent, G., W. Amara, J. Mansourati, O. Bizeau, P. Couderc, N. Delarche, ... EDUCAT registry investigators. 2014. Role of patient education in the perception and acceptance of home monitoring after recent implantation of cardioverter defibrillators: the EDUCAT study. Archives of Cardiovascular Diseases. 107(10): 508–518.

O'Halloran, P.D., L.J. Murray, G.W. Cran et al. 2005. The effect of type of hip protector and resident characteristics on adherence to use of hip protectors in nursing and residential homesean exploratory study. Int. J. Nurs. Stud. 42: 387–397.

Orlikwski, W. and D. Robey. 1991. Information technology and the structuring of organizations. Information Systems Research. 2: 2, 143–169.

Peek, S.T.M., E.J.M. Wouters, J. van Hoof, K.G. Luijkx, H.R. Boeije and H.J.M. Vrijhoef. 2014. Factors influencing acceptance of technology for aging in place: A systematic review. International Journal of Medical Informatics. 83(4): 235–248.

Pino, M., C. Granata, G. Legouverneur, M. Boulay and A.-S. Rigaud. 2012. Assessing design features of a graphical user interface for a social assistive robot for older adults with cognitive impairment. Gerontechnology. 11(2): 383.

Taverner-Smith, K. and G. DeVet. 2006. Further exploring hip protector use. Australasian Journal on Ageing. 25: 215–217.

Vaujauny, F.X. 2006. Pour une théorie de l'appropriation des outils de gestion: vers un dépassement de l'opposition conception-usage. Revue Management et Avenir. 9: 109–126.

Wettstein, M., H.-W. Wahl, N. Shoval, G. Auslander, F. Oswald and J. Heinik. 2014. Cognitive status moderates the relationship between out-of-home behaviour (OOHB), environmental mastery and affect. Archives of Gerontology and Geriatrics. 59(1): 113–121.

8

Monitoring Medication Adherence in Smart Environments in the Context of Patient Self-management

A Knowledge-driven Approach

Patrice C. Roy,[1,a,*] *Samina Raza Abidi*[2] and *Syed Sibte Raza Abidi*[1,b]

1. Introduction

Ambient Assisted Living (AAL) technologies, such as smart environments, are suitable for assisting individuals to self-manage their health in a home-based setting (Roy et al. 2011). The primary function of AAL technologies is to provide adequate and relevant support at the opportune moment.

[1] NICHE Research Group, Faculty of Computer Science, Dalhousie University, 6050 University Avenue, PO BOX 15000; Halifax, NS B3H 4R2, Canada.
[a] Email: patrice.roy@dal.ca
[b] Email: sraza@cs.dal.ca
[2] Medical Informatics, Faculty of Medicine, Dalhousie University, Halifax, NS B3H 4R2, Canada. Email: samina.abidi@dal.ca
* Corresponding author

From a healthcare perspective this support can be in terms of monitoring the home-based activities of patients with the intent to remind and guide them about impending healthcare tasks. Ambient services, such as activity recognition, provide the functionality to establish the situational context of an individual and then provide context-sensitive self-management support to the specific individualized needs of the patient.

Lifetime healthcare is an emerging health paradigm, where the focus is on assisting patients to achieve health targets and avoid harmful lifecycle choices in order to lead a healthy life. Disease self-management is an important aspect of lifetime healthcare, where the intent is to engage and empower the patients to self-manage their condition by adhering to their therapeutic plan (such as taking medications regularly), maintaining a healthy lifestyle, performing health tasks (such as follow-up visits, investigations) in a timely manner and mitigating risk factors. Self-management programs guide and motivate patients to achieve self-efficacy in self-management of their disease through a regime of educational and behavioral modification strategies.

In the context of lifetime healthcare, where patients are required to self-manage their condition, it is imperative that they follow the prescribed therapy, i.e. adhere to their medication plan by taking their prescribed medications in the right dose and at the right time. Medication adherence is defined as "the extent to which patients follow the instructions they are given for prescribed treatments" (Haynes et al. 2002). Sub-optimal medication adherence among patients, particularly among those with chronic diseases and receiving long-term therapies, is a problematic issue leading to serious healthcare costs and discomfort for the patient (Osterberg and Blaschke 2005; WHO 2003). Studies have shown that around 20–30 per cent of all medication prescriptions are never filled and about 50 per cent of medications for chronic diseases are not taken as prescribed (DiMatteo 2004; Haynes et al. 2008; Peterson et al. 2003). This trend is particularly acute in patients with cardiovascular diseases and asthma, where non-adherence rates are 50 per cent for cardiovascular patients (Kronish and Ye 2013), and between 30–70 per cent for asthma patients (Bender et al. 1997), with < 50 per cent of children adhering to their prescribed inhaled medication regimen (Milgrom et al. 1996). Consequences of non-adherence to medications are profound both healthwise and economically—healthcare costs attributed to medication non-adherence are estimated to be between $100 and $300 billion annually in US, representing 3–10 per cent of total US healthcare costs (Aitken and Valkova 2013). The majority of costs are attributed to (a) increase in service utilization at physicians' clinics, emergency department, nursing homes, hospices, etc.; (b) therapy augmentation due to disease complication or progression, and/or comorbidity development that results in increased pharmaceutical costs; (c) avoidable diagnostic lab and radiology tests and procedures that are warranted due to worsening of the condition (Aitken and Valkova 2013). Most of the studies on the burden of non-adherence typically estimate direct healthcare costs and do not include productivity loss, such as progressive functional disability, reduced work capacity and premature retirement. As a result, the real burden of non-adherence is believed to be higher than the estimated costs.

A number of strategies to promote medication adherence have been presented in literature (Atreja et al. 2005). The most significant of these construct include: (i) simplifying the drug regimen, such as reducing drug intake frequency (Morningstar et al. 2002), using simple language to explain regimen (Eraker et al. 1984) and using adherence aids such as pill-boxes (Cramer 1998); (ii) providing simple and effective patient education in order to enhance patient's understanding of the drug regimen and health consequences of non-adherence (Burgoon et al. 1987; Daltroy et al. 1991; Hall et al. 1988; Katz 1997); (iii) improving communication with the patient and if necessary, involving patient's families in discussions regarding the regimen and building trust in patient-provider relationship (Morrow 1997; Rosenberg et al. 1997); (iv) helping patient modify behaviors that contribute to poor medication adherence by addressing patient's beliefs, intensions, confidence, social support, environment, etc. (Roter et al. 1998). Although the above mentioned adherence interventions have proven to be effective to a varying degree, almost half of the interventions seem to fail (Haynes et al. 2008). One of the reasons of sub-optimal outcomes of these interventions is the inability to accurately ascertain whether the patients are taking their medications (most approaches rely on self-reported data and prescription refills, which are unreliable accounts of medication adherence). We argue that one approach to improve medication adherence is to unobtrusively monitor the patient's intake of medications at the right time and to remind them in case they are non-compliant to their medication regime.

We posit that AAL technologies in smart environments can help patients through personalized self-management assistive acts by (a) monitoring a patient's compliance to recommended healthcare activities (such as medication adherence, daily exercise, etc.), and (b) in the event that the patient is forgetful of the recommended activities, reminding the patient to perform the self-management activity (via smartphone local notification, smart home speaker) and recording the completion of the self-management activity. Thus AAL technologies can potentially contribute to the effectiveness and efficacy of self-management programs (Patterson et al. 2015).

Self-management programs focus on improving the patients ability and motivation to comply with their therapy plan by helping them overcome societal and behaviour barriers, such as unhealthy diet, stress, medication non-adherence, sedentary lifestyle and so on, that compromise their ability to self-manage their disease. Self-management programs aim to improve the efficacy of the patient to achieve the therapeutic targets set by their care providers by providing focused educational and motivational support. An important aspect of helping a patient to achieve efficacy in self-managing their condition is to help him/her overcome any perceived barriers that may limit the patient's ability to self-manage his/her condition. For instance, an important element of self-management is medication adherence as that allows the patient to achieve a state of well-being and disease control. The barrier to medication adherence is that patients at times either forget or choose to ignore taking their medications (due to some misconceptions) which affects their wellbeing. Hence, to promote efficacy, self-management programs tend to focus on reminding, educating and motivating patients to take their medications on time.

Medication non-adherence is a common behaviour amongst chronic disease patients and mostly because patients forget to take their entire medication regime. Assistive technologies can play a significant role in disease self-management by detecting when a patient has taken his/her medication as per the medication schedule, reminding the patient to be compliant when forgetful of taking their medications and educating and motivating them to be more attentive to their medication schedules. To meet this end, we need a knowledge-centric self-management system that comprises (a) assistive technologies that understand the environment and can detect medication-related events; (b) information about the patient's medication schedules and preferences; (c) knowledge of the working of the self-management strategy and the corresponding medication adherence educational content; and (d) an up-to-date understanding of the patient's medication compliance behaviour and daily activities. We argue that the application of semantic web technologies in conjunction with assistive technologies provide a unique opportunity to formally represent and reason over the context, events and responses in order to design personalized medication adherence self-management programs. The ability to represent and reason over knowledge will render the functionality to infer high-level contextual information (e.g. activity) from smart environment events (low-level) and then to direct self-management support recommendations/actions to patients with respect to their context and health needs. Furthermore, abstraction of the patient's medication adherence practices can be used to intelligently develop more effective self-management strategies that will include more focused, timely and impactful educational and motivational messages through the smart environment's assistance services.

The remainder of the chapter is organized as follows: firstly, an overview of activity recognition is presented, with emphasis on ontological-driven activity-recognition approaches; secondly, an overview self-management program and of a previous ontology-driven self-management program (Abidi and Abidi 2013) is given. The next section presents how a self-management program could interact with a smart environment. Furthermore, this section also presents an event model that can be used in the context of self-management of disease with a smart environment. Finally, we conclude the chapter, by mentioning the future prospects.

2. An Overview of Activity Recognition in Healthcare

Activity recognition can be described as a process of inferring an agent's ongoing tasks from the observed environmental changes triggered by the agent's behaviour (Roy et al. 2011). Activity recognition approaches can be classified according to four criteria:

- What are the activity recognition objectives?
- What is the relation between the observed agent and the activity-recognition component?
- How are the activities monitored?
- How are the activities modelled, represented and processed to infer the ongoing tasks?

The activity recognition objectives concern the output of the inference process. For instance, we may only be interested in the observed agent's goals (intent recognition) (Horvitz et al. 1998). We may also be interested in how the ongoing activities are carried out (plan recognition) (Roy et al. 2009). In plan recognition, the objective is to infer the observed agent's plan and objectives. A plan is a partially ordered set of actions/tasks that must be carried out in order to complete some goals. Thus, the objective is to evaluate the hypotheses about the composition of the plan structure and the associated goals. Finally, we may only be interested on the current activities without any information on the activities' states (e.g. current step).

The relation between the observed agent and the activity recognition process can be divided into three distinct categories (Geib 2006). The first category is keyhole recognition, where the observed agent does not attempt to influence or mislead the recognition process (Cohen et al. 1982). Thus, the observed agent carries out activities without taking into account the fact that it is observed (no cooperation). The second category is intended recognition, where the observed agent adapts its behaviour in order to help the recognition process (Kautz 1991). Thus, the observed agent can receive clarification request from the recognition process (positive cooperation). The third category is adversarial recognition (Geib 2006), where the observed agent is opposed to the observation of its behaviour. Thus, the observed agent tries to abort the recognition process by carrying out actions by hiding its intention (negative cooperation).

Activities can be monitored by several sources, such as vision-based (Bilinski and Bremond 2011), sensor-based (Tapia et al. 2004) and self-reported-based (Bardram et al. 2012; Haakstad et al. 2010). Vision-based approaches use computer vision techniques to analyse visual information in order to obtain features that will be used in order to infer the current activities. They often use probabilistic graphical models to learn and infer the activities (Weinland et al. 2011). The main issue with this type of recognition is the intrusiveness of video sensing equipment and privacy concerns of all observed actors (patient, relatives, caregiver, etc.). Sensor-based approaches are often used in smart environments, such as smart homes (Cook and Krishnan 2014). These approaches can use several types of technologies, such as environmental sensors (passive infrared (PIR) sensors, radio-frequency identification (RFID) tags, flow meters, etc.), wearable sensors (Chan et al. 2012) and smartphones (Guidoux et al. 2014).

Activity modelling, representation and inference approaches can be classified into two main categories: data-driven and knowledge-driven activity recognition. Data-driven activity recognition uses machine learning techniques in order to obtain activity models from datasets (Hu et al. 2014). These datasets contain sensor data that are usually annotated with activity labels (Ye et al. 2013). Activity inference is usually based on probabilistic and statistical reasoning, such as probabilistic graphical models. For instance, Hidden Markov Models (HMM) (Rabiner 1989) and Dynamic Bayesian Networks (DBN) (Friedman et al. 1998) can be used to infer activities from manipulated objects (RFID tags on objects and RFID antenna on gloves) (Patterson et al. 2005). Other techniques have been investigated, such as naïve Bayes (Phua et al. 2009), Support Vector Machine (SVM) (Kadouche et al. 2010), Conditional Random Field (CRF) (Nazerfard et al. 2010), Decision Tree (Stankovski and Trnkoczy 2006), Artificial Neural Network (ANN) (Delachaux et al. 2013) and nearest neighbours (Gillani Fahad et al. 2015). Data-driven activity-recognition approaches are considered better in handling noisy, uncertain and incomplete data (Okeyo et al. 2014). Furthermore, data-driven activity-recognition approaches, such as HMM, DBN and CRF, usually support modelling and inference of complex activity patterns (interleaved activities) (Okeyo et al. 2014). However, the main drawback of data-driven approaches is that large amounts of initial training data are required in order to learn the activity models. Since the same user can carry out activities in different ways, all variants must be present in the dataset in order to learn properly the activity models. Also, since the user can change its behaviour, the learnt models can be less effective and need to be adapted/retrained. Finally, since different users don't share the same behaviour, it is difficult to reuse the same models between users.

Knowledge-driven activity recognition is based on logical modelling and reasoning (Chen et al. 2008). It uses knowledge representation formalisms to model activities and sensor data and then use logical reasoning for activity inference. Logical-based approaches, such as those by Kautz (1991) and Christensen (2002), develop a theory using first order logic for formalizing the recognition process into a deduction or an abduction process. Since knowledge-driven approaches are grounded in logic theory, they are semantically clear in modelling and representation and elegant in inference and reasoning (Chen et al. 2012). However, lack of support for uncertainty reasoning and temporal knowledge pose challenging issues in order to reason about complex activities and uncertainties related to sensor and inferred data (Helaoui et al. 2013). For instance, uncertainty from sensor and inferred data needs to be evaluated, based on some characteristics (e.g. sensor properties), to be semantically modelled and to be propagated through semantic reasoning (Aloulou et al. 2015). Recently, several approaches use ontologies for modelling and representing activities. The following section presents an overview of these ontological-driven approaches.

Ontological-Driven Activity Recognition Approaches

Ontological-driven activity recognition can be carried out either using ontology reasoning or ontologies with rules (Meditskos et al. 2013b). Approaches based on ontology reasoning use, for instance, Description Logics (DL) (Baader et al. 2004) to model contextual elements (e.g. events, activities, users, locations, time) and standard DL inference services to derive logically implied context information (Chen et al. 2012). Approaches based on ontologies with rules use, for instance, Web Ontology Language (OWL) (Horrocks et al. 2007) to represent activity-related information (contextual) and SPARQL Inferencing Notation (SPIN) (Knublauch et al. 2011) to derive high-level contextual activity information (Meditskos et al. 2013b). The remaining part of this section presents some recent ontological-driven activity-recognition approaches.

Chen et al. model and its extensions

In (Chen et al. 2012), the authors present an activity-recognition approach based on context and activity ontologies and use standard semantic reasoning and classification. Smart home entities (e.g. device, furniture, location, time, sensor) and their interrelationships are described with the context ontology. For instance, the sensor model can capture domain knowledge, such as a contact sensor attached to a teapot in the second cupboard to the left of the sink in the kitchen. Thus, it is possible to infer the corresponding objects and location from the activation of the sensor. For instance, if we observe an activation of the previous sensor, it implies that the user performs an activity in the inferred location (kitchen) with the inferred object (teapot). It should be noted that the sensor activations are interpreted as user-object interactions, ignoring how the object is used and when it is deactivated. Thus, a user-object interaction is equivalent to instantaneous sensor activation and is interpreted as an object that has been used for carrying out an activity.

In this model, the segmentation approach for extracting an observed situation at a specific time points to aggregate individual user-object interactions according to a specific time window size. The activity models are described with several properties, which can be categorized into three groups. The first group represents contextual information (time, location, actors, resources) within which the activity takes place. The second group represents causal and functional relations (conditions, effects, goals, duration) that are used for activity reasoning. The third group represents type and interrelationship between activities, allowing representation of complex activity type, such as Activity of Daily Living (ADL) (Katz et al. 1963), at different levels of abstraction. The activity-recognition algorithm uses a generated situation (aggregated sensor observations) at a specific time point and semantic reasoning services in order to infer the activity from the situation properties. For continuous activity-recognition, the proposed model uses the time window technique (sliding or at opportune times). The time window is used for sensor activation fusion, which generates a situation. However, the proposed model doesn't take into account uncertainty. This model was extended in order to take into account some limitations related to continuous activity recognition and temporal information.

In (Okeyo et al. 2014), the authors improve the real-time continuous recognition component by proposing a dynamic sensor data segmentation. The dynamic segmentation model is based on varied time windows, which can expand and shrink the window size according to temporal information of sensor data and activities and the state of activity recognition. Activity recognition is triggered according to three modes. In the first mode, activity recognition is performed each time a sensor is activated. In the second mode, activity recognition occurs periodically at regular intervals during the length of the time window. The activity recognition engine must check if any new sensor activation was observed before starting the reasoning process. In the third mode, activity recognition occurs only at the expiry of the time window (no dynamic sizing). When the time window is deactivated, all sensor activations used within it are discarded. If a specific activity is inferred, the time window is truncated if all the properties required to describe the activity is asserted, and is expended if more time is required to obtain missing properties (must not go beyond maximum activity duration property). If no specific activity is inferred, the current inferred generic activity is used to expend the time window (must not go beyond the duration of its subclass activity that has the longest duration).

In (Okeyo et al. 2014), the authors extend the activity model with temporal concepts, such as the Allen interval relations (Allen and Ferguson 1994), in order to model composite activity patterns. Composite activities consist of simple activities (e.g. making tea, preparing pasta, watching television) that are carried out in a sequential way (activities do not overlap) or interleaved/ concurrent way (activities overlap). For instance, the user can make pasta and watch television in an interleaved way, or can wash her hands and then make tea (sequential). The composite activity recognition is based on a rule-based engine. The results of the (simple) activity recognition are aggregated and the rules infer qualitative temporal relationships between them and derive corresponding composite activities.

Helaoui et al. model

In (Helaoui et al. 2013), the authors present a probabilistic ontological framework in order to recognize multilevel activities. It is based on log-linear description logics (Niepert et al. 2011), which support modelling and reasoning with uncertainty by combining description logics with log-linear models. In log-linear description logics, uncertain axioms have weight values, where high weights indicate high confidence that the related axioms hold. The main inference task is the maximum a-posteriori (MAP) query: *Given a log-linear ontology, what is a most probable coherent ontology over the same class and property names?*

In the proposed model, activities are divided into four levels. The first level consists of atomic gestures, such as *reach dishwasher door*, where actions cannot be decomposed to simpler ones and are often considered as instantaneous events (very short duration). The second level consists of manipulative gestures, such as *close dishwasher*, which are composed of atomic gestures (*reach dishwasher door*, *push dishwasher door* and *close dishwasher door*) and have a short duration (few seconds). The third level consists of simple activities, such as *put tableware in dishwasher*, which are characterized by temporal sequences of manipulative gestures (*fetch an object*, *open the dishwasher*, *put down the object* and *close the dishwasher*) and have short duration (few seconds). The same simple activity can be characterized by different sequences of manipulative gestures. The fourth level consists of complex activities, such as *clean up*, which are characterized by concurrent execution of simple activities (*put tableware in dishwasher* and *clean table*) and can last from a few minutes to a few hours.

Recognition of atomic gesture is based on supervised machine learning. Body-worn sensors (accelerometers, gyroscope) are used to infer movements, simple actions and body posture. RFID tags are used to infer objects usage. For instance, the atomic gesture, *pushing a door*, is inferred by matching the used object (door) with the inferred body action (pushing). Recognition of manipulative gestures is based on the inferred atomic gestures during a time window (for instance, one second) and the semantic structure of activities. Inconsistencies (incompatible inferred manipulative gestures) are resolved by computing the most probable consistent ontology and then use on it the standard DL reasoning to infer manipulative gestures. Recognition of simple activities is based on manipulative gesture occurrences during a time window (for instance 4 seconds). In order to support a simple form of temporal reasoning, the sequence of manipulative gestures is represented as a list of triadic properties (a manipulative gesture, its actors and its order in the time window). These triadic properties are represented in DL as binary properties by using a representation pattern (Hayes et al. 2006). The recognition process computes the most probable consistent ontology and uses on it DL reasoning to derive a set of simple activities. Recognition of complex activities is based on the inferred simple activities during a time window (for instance, 30 seconds). The reasoning process is similar to the one used in manipulative gesture recognition. A major limitation of this proposed approach is the simplistic temporal modelling, which prevents recognition of fine-grained activities with rich temporal intra-relationships.

Aloulou et al. model

In (Aloulou et al. 2015), the authors propose an ontological-driven recognition approach based on semantic rules and the Dempster-Shafer theory (Shafer 1976). A semantic model represents the knowledge about entities in the environment (users, activities, locations, assistive services, sensors and devices) and the relations between them. The uncertainty level of contextual information (e.g. user location) is based on hardware characteristics and operational parameters of the sensor used to obtain the information. The levels of certainty received from the sensing layers are propagated through the reasoning of higher-level context information (e.g. activity) by including uncertainty information in the semantic rules. For instance, if sensor *L1* is deployed in the living room and its state is *ON* with a certainty level of 80 per cent, then the patient is detected in the living room with a certainty level of 80 per cent. In order to resolve conflicting situations (e.g. uncertainty emerging from different sources), the Dempster-Shafer theory is applied to obtain the most realistic and consensual uncertainty value.

However, this approach provides more accurate results when it uses highly coupled activity-related sensors. If a highly coupled sensor fire (event) occurs, then one specific activity is detected. The implemented system provides exact detection only when highly coupled sensors are used (with or without other low coupled sensors). In the real context, sensors are often lowly coupled, where sensor events can be produced by several activities.

Roy et al. model

In (Roy et al. 2011), the authors propose activity-recognition approaches based on description logics and the possibility theory (Dubois and Prade 2006). Possibility theory is an uncertainty theory dedicated to process incomplete information and is based on a pair of dual set functions (possibility and necessity measures). It allows to model, in a more flexible way than probability theory, uncertain judgment (partial belief) from domain experts, which is incomplete and imperfect. The activity recognition process can be separated into four layers.

The event manager (layer 1) collects sensor events and evaluates whether the patient carried out an action. However, it should be noted that this proposed model assumes that this layer already exists. Context recognition (layer 2) infers the plausible low-level contexts (smart home environment's state and patient profile) from smart home ontology and information retrieved from sensor events (observation). A low-level context is a set of properties that are shared by some states in the environment's state space (a context is subset of state space). Since an observation is partial, it is possible to have multiple contexts that explain the current observed state. Contexts are partially ordered in the ontology with the context subsumption relation, which is based on concept subsumption relation from description logics (Baader et al. 2004).

Action recognition (layer 3) infers the most plausible low-level action that was carried out according to the set of plausible contexts from the current and previous observations. It uses an action model where each action is defined by a set of context transitions. Each transition represents a plausible transition between two contexts when the action is carried out. Each transition is quantified with a possibility value, indicating the possibility that if we observe a specific context transition, then it results from carrying out the action related to the transition. Thus, we obtain a set of plausible actions, where each action is assigned a possibility and necessity measures according to the plausible context transitions. The inferred action is the one with the highest possibility and necessity measures. If several actions are selected, then the recognition process uses the action ontology and its action subsumption relation to select the most specific among the ones that generalize all selected actions. For instance, if the selected actions are *open cold tap* and *open hot tap*, then the inferred action is *open tap*, since it is more specific than *any action*.

Behaviour recognition (layer 4) infers hypotheses about the plausible high-level user's behaviour related to the erroneous or coherent accomplishment of some intended activities. Each activity is described with an activity plan structure, where actions are partially ordered with a temporal relation and possibility distributions on the context set, indicating the possibility that the current observed context is related to a coherent or erroneous realization of the activity. The temporal relation between two actions indicates the possible duration between the two actions' realizations (possibility distribution on duration time). According to the sequence of observed actions (inferred from layer 3) and activity plans, a set of plausible activities is selected (activities that have a plausible partial/complete realization path). A realization path is a subset of the observed action sequence where each action is associated with an action in an activity plan, thus having a coherent partial/complete realization of the activity. A behavioural hypothesis is a subset of plausible activities that are carried out in a coherent way (completed or not), while other activities, if any, are carried out in an erroneous way. For each hypothesis, the possibility and necessity measures related to coherent and erroneous behaviours are evaluated according to the selected activities' models and partial paths. Possibility and necessity measures allow selection of the most plausible hypotheses.

This proposed model has some limitations. For instance, the uncertainty from the sensor data is not taken into account (only the action and activity models). Defining the actions and activities require a lot of information (context/action transitions), which can be alleviated by using machine learning to infer the model parameters. Finally, the behaviour recognition (layer 4) could be optimized, since it is possible to obtain several realization paths for the same activity.

Meditskos et al. model

In (Meditskos et al. 2013b), the authors propose an ontology-driven recognition approach based on OWL reasoning and SPARQL rules (*MetaQ*). The proposed framework consists of three layers. The *representation layer* provides the ontology for modelling basic activity-related information (activity classes, actors, locations) and two high-level activity correlations (classifications and compositions properties). Each atomic and complex activity is represented as an instance of the activity class and has properties. Each activity has a time range (start time and end time), actors (persons carrying out the activity), participants (physical entities, such as objects and other persons participating in the activity), spatial information (area where the activity is carrying out) and correlation properties (sub-activities and classifiers).

The *interpretation layer* infers complex situations through an iterative combination of OWL reasoning (first step) and SPARQL queries (second step). Complex activities are obtained by aggregating and interpreting primitive activities, using SPARQL construct queries. The interpretation queries (SPARQL) help, for instance, to infer the location of an activity based on the detected objects. The first category of queries, *classification,* allows inferring complex activity from the presence of its dependencies. For instance, the *bed exit* activity is inferred from the presence of *out of bed* and *night sleep* activities' instances. The second category of queries, *composition,* allows inferring composite activities from its sub-activities. For instance, a *nocturnal* activity is detected when its sub-activities—*bed exit* and *in bathroom*—are observed.

The *activity meta-knowledge layer* encapsulates a formal description of the interpretation semantics by using OWL activity meta-patterns (Meditskos et al. 2013a). The activity pattern ontology is based on the DnS pattern of DUL (Gangemi and Mika 2003), which serves as a metamodel for capturing structural aspects of atomic and compound activities. The situation component of the metamodel provides an abstract description of the complex activity in terms of the domain activity types that are involved. The description component of the metamodel can be seen as a descriptive context that further classifies the activity classes of the situation, creating a view. This metamodel is used to describe the classification and composition patterns. By using these patterns, it is possible to generate the composition and classification queries.

However, this proposed model has some limitations. The current version doesn't take into account uncertainty. Also, activity recognition is done offline. Finally, the time window length and the triggering criteria for the observation collection process are still under investigation (Meditskos et al. 2015).

3. Self-Management Programs for Chronic Disease Management

Self-management programs are designed to help individuals with health issues to get engaged in their care process and play an active role in managing their condition (Quinn et al. 2015). Self-management can be defined as the person's ability to manage the symptoms, physical and psychological consequences, treatment and lifestyle changes inherent in living with a chronic health condition. Self-management programs involve a dynamic and continuous process of self-regulation with the intention to enhance an individual's efficacy to self-manage their condition. Self-management programs recommend strategies that will help the individual develop a degree of self-efficacy, motivation and knowledge to take effective control over his life with a chronic illness (Quinn et al. 2015). The key element of self-management programs is to induce positive behaviour modification amongst individuals so that they are more inclined and confident in adhering to their therapy plan. In case individuals are unable to adhere to their therapy, then plan self-management programs that recommend behaviour modification strategies to overcome perceived barriers/challenges faced by the individual in adhering to the prescribed therapy plan. In this case, problem solving skills are introduced to the individual and through a set of guided tasks and personalized messages, the individual tries to achieve self-efficacy in managing his/her condition (Bodenheimer et al. 2002). Of course, the success of the self-management program depends on timely responses to the individual's actions and the relevance of the motivation message/educational content with respect to the individual's content. We argue that assistive technologies can play a significant role in capturing the individuals' context and monitoring their actions. And, when this information is coupled with a knowledge-centric self-management program, there are unique opportunities to not just monitor the individual but also to guide with context-sensitive, personalized and evidence-based behaviour modification recommendations that will help towards self-management of the condition.

The most common theoretical basis for self-management is the Social Cognitive Theory (SCT) (Bandura 2001). SCT provides a theoretical framework for developing interventions to modify a variety of health behaviours. Use of motivational enhancements can lead to behaviour modification (Bandura 2004). An educational intervention, which is designed to provide information, behavioural strategies and incentives to motivate individuals to meet health-specific goals, has two objectives. The first objective is to help the individual to modify his behaviour. The second objective is to enhance the individual's perceived self-efficacy and outcome expectation. Self-efficacy refers to a person's belief that he can successfully carry out a specific action in a particular situation. Thus, in order to ensure permanent behavior change, SCT aims to encourage self-regulation and self-efficacy. According to SCT, a patient's behavior is affected by expectation of self-efficacy (belief that he can carry out successfully a necessary behavior) and outcome expectations (belief that a behavior will produce the desired outcome). Self-efficacy, outcome expectations and cognitive skills are developed and modified by social and environmental influences (social environment, encouragement and persuasions from credible people, observing behaviour of other people, making judgments if the desired outcomes was obtained) (Clark and Dodge 1999). Self-management programs based on SCT demonstrate positive effects on health behavior in several chronic diseases, such as heart disease (Oka et al. 2005), diabetes (Aljasem et al. 2001), and on stopping negative behavior, such as smoking and sedentary lifestyle (Boudreaux et al. 2003).

Ontology-driven Self-Management Framework

Computerized self-management programs have been used to influence behavioral changes to help people make informed choices about therapeutic options, risk assessments and lifestyle modifications. For instance, Abidi and Abidi (2013) proposed an ontology-driven personalization framework for designing self-management interventions based on SCT.

In agreement with SCT, this framework engages patients (a) to identify their perceived barriers to reach their self-management targets and to select behavior modification strategies to overcome the identified perceived barriers. In accordance with the patient-identified barriers to self-management, the framework delivers personalized behavior modification and self-efficacy fulfilment interventions (behavioral change strategies, motivational messages, educational material) that specifically target the patient's self-selected behavior modification goals and the patient's profile characterizing health apart from, behavioral and psychosocial aspects. The use of mobile devices to monitor and educate the patients allow ubiquitous interventions (Abidi et al. 2013).

The patient's *health profile* is based on a set of validated health assessment tools to determine his health status in accordance with a set of disease-specific parameters that are used to personalize health and behavioral change interventions. The assessments are usually questionnaires; for instance, the health profile for cardiac patients can be based on levels of stress, lifestyle, blood pressure and heart rate. The patient's *behavior profile* highlights the behavioral and psychological disposition of a patient towards disease health management and is based on evidence-driven behavior models (health belief model, SCT, motivational interviewing). This profile allows tailoring of self-management programs in order to comply with the interests, needs and challenges of the patient. For instance, the behavior profile for managing cardiovascular risks is based on SCT markers (self-efficacy, self-regulation, social support, outcome expectation and socio-economic status).

The identified *barriers* are the ones that a patient perceives to be factors in his/her self-management of the disease. For instance, the patient can identify several barriers to medication adherence (drug regimen complexity, cost of drugs, forget medication time, forget refilling medication, unable to manage stress), regular exercise and healthy lifestyle. In order to counter these identified barriers, the patient identifies *self-efficacy goals* that he will like to achieve. For instance, the patient wants to comply with his drug plan. Once the patient selects a set of personal self-efficacy goals, he selects specific strategies that he will adopt to achieve behavioral self-efficacy and self-regulation as a means to overcome his barriers that are hindering the self-management of his health condition. The available strategies, which are action plans, are shortlisted on the basis of the patient profile and identified barriers.

The patient's profile and barriers to overcome are used to personalize self-management strategies. The self-management strategies are intended to assist the patient to modify his behavior in order to reach the self-efficacy goals, thus achieving self-efficacy in managing the disease, which leads to achieving positive health outcomes. Usually, the patient follows a self-management improvement strategy for a specific duration (for instance, two to four weeks). During this intervention period, educational and motivational messages are proactively delivered to the patient in order to help him overcome the selected barriers. Furthermore, the patient's compliance to the self-management plan is monitored.

At the end of the intervention period, a reassessment (based on validated assessment tools) of the patient's self-efficacy to successfully self-manage his disease is carried out. Depending on the assessment results, the patient can select a new barrier to overcome or continue to reinforce the current one.

This proposed framework uses an ontology to (a) model the theoretical framework SCT in terms of SCT concepts; (b) model the health assessment tools to develop the patient's profile; (c) instantiate the self-management strategies and related educational messages; and (d) model the personalization rules that integrate the health and SCT models with corresponding educational and motivational messages. Figure 1 presents an overview of the high-level concepts for modelling self-efficacy strategies based on SCT.

The *patient profile* (health and SCT-based behavior profiles) modelled by the ontology is used to design personalized self-efficacy strategies. This approach identifies a self-efficacy *goal*, finds out the *barriers* to the achievement of this goal and then carries out a *strategy* in order to follow

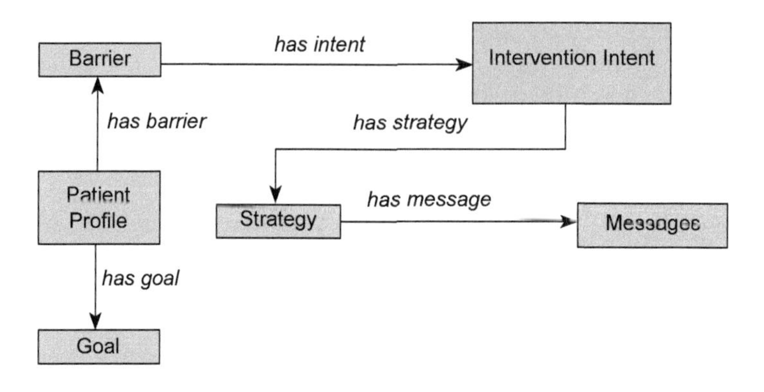

Fig. 1. High-level view of concepts used in ontological modelling of self-efficacy strategies based on SCT.

the realization of this goal. *Goal* represents the self-management objective that the patient plans to achieve through the self-efficacy interventions. For instance, typical goals are compliance to medication, healthy lifestyle and stress management. *Barriers* represent the perceived barriers to the achievement of the patient's goals. For instance, typical barriers that hinder the goals are low motivation, lack of knowledge, busy lifestyle, financial considerations and forgetting the task schedule (e.g. taking medication, refilling medication). *Intervention intents* represents specific focus areas that are based on barriers. For instance, *improved medication intake on time* and *improved skills to deal with medication regimen complexity* are intervention intents for barriers related to the *medication adherence* goal. The intervention intents selected by the patient are pursued through a *strategy*, which represents a concerted plan to overcome a barrier. Each strategy comprises a set of actions with accompanying motivational and educational messages and targets to be achieved during a specific timeframe. The hierarchy of strategies allows tailoring of the strategy *message* according to the patient's profile. For instance, the high-level *strategy for remembering to take medication at proper time* can have sub-classes, where each one represents a set of strategies that can be offered to the patient in order to deal with the '*taking medication on time*' issue. Each strategy *message* is associated with a set of constraints in the patient's profile. If the constraints are satisfied, then the message is judged relevant to a patient. This self-management framework was applied to design personalized self-management strategies for cardiac risk factors, delivered to the patient through a mobile app (Abidi et al. 2013).

4. Smart Environments and Self-Management Program

Given the availability of an ontology-based self-management framework (Abidi and Abidi 2013), in this section we discuss how assistive technologies can be integrated with knowledge-centric self-management programs that operate within smart environments, such as smart homes and smartphones. The advantage of the said approach is the ability to infer new clinical facts from real-time monitoring services based on ambient and wearable sensors (Roy et al. 2014) and respond with contextualized self-management interventions/messages/actions. These clinical facts can be used to assess health and behavior information related to the patient's self-management profile and to assess the patient's compliance with the personalized self-management plan. Such monitoring services allow the retrieving of relevant self-management information in a less intrusive way. Furthermore, self-management program strategies can use smart environments in order to deliver motivational and educational messages to the patient. For instance, the smart home effectors (speakers, TVs, tactile screens) can be used to notify the patient in accordance with the patient profile (self-management and smart environment) and preferences.

Figure 2 shows interactions between components of a computerized self-management program and a smart environment for the self-management goal *medication adherence*. The self-management program is based in the ontology-driven self-management framework. The smart environment's components are related to activity recognition and monitoring. Behavior modelling is used for modelling the patient's profile (health, SCT-behaviour, activity pattern, etc.). In this example, the patient selects the *medication adherence* goal and wants to tackle the '*forget to take medication on time*' barrier with the help of strategies related to the '*improve medication intake on time*' intervention intent. The self-management program creates a personalized self-management plan based on the patient's profile and selected goal, barrier and intervention intent. This plan (strategies) is designed to assist the patient to modify his behavior and achieve (outcome) positive self-efficacy and self-regulation in self-managing the disease. This self-management improvement plan is carried out for a specified period (for instance, one month). During this intervention period, motivational messages reinforcing the '*take medication*' task are proactively delivered to the patient in order to achieve self-efficacy to overcome the barrier. It should be noted that the message can be delivered through the smart home or smartphone. In order to monitor the compliance with the self-management plan and messages delivered during the intervention period, events from the smart environment's decision support system can be used

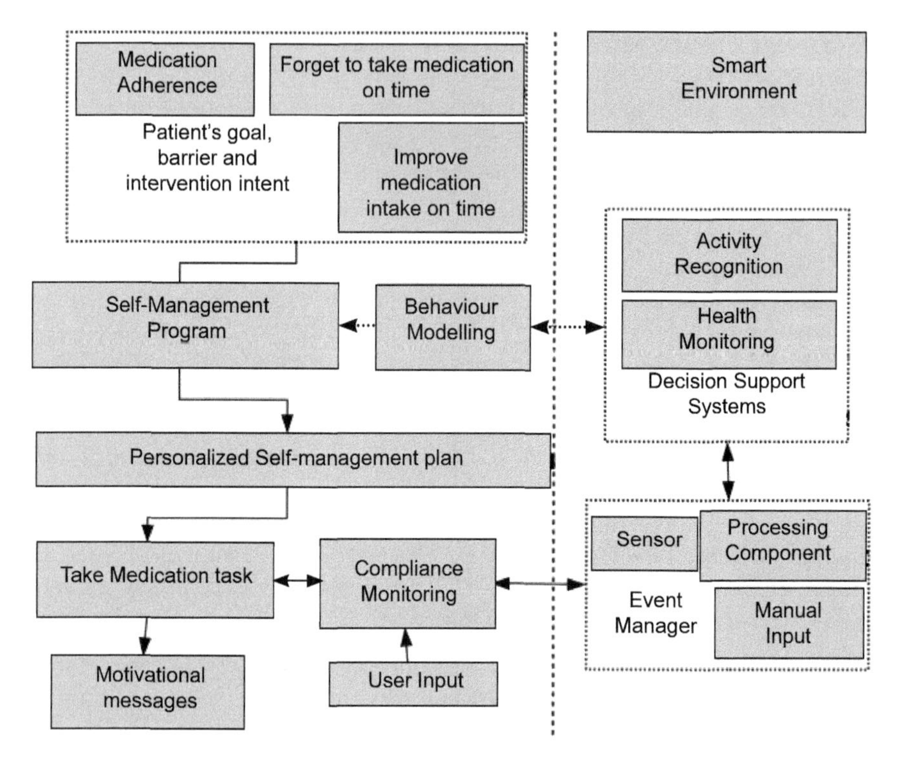

Fig. 2. Interactions between components of self-management program and smart environment.

as input, instead of using only the user input. It should be possible to adapt the personalized self-management plan depending on the events received by the compliance monitoring. Smart environment's events can come from several sources, such as sensors (accelerometer), manual input (computerized questionnaire) and processing components (decision support systems, activity recognition, heath-risk monitoring).

Smart environment's events are usually restricted to sensor-based observations (Ye et al. 2013). However, in a self-management context, events can come from the processing components (activity recognition, fall detection, health monitoring) and manual input (health assessment questionnaire). In order to interpret and fuse events into the high-level domain-specific information used in components, such as medication-adherence monitoring and activity recognition, events are modelled based on an event taxonomy.

Representing Events

Events are often represented by a tuple indicating the observation timestamp, the observation source (sensor ID) and the observation value (e.g. sensor state). Several approaches proposed ontologies for representing the events; for instance, Compton et al. (2012) proposed an ontology for sensor networks. However, the proposed ontologies often neglect uncertainty and other event sources. Figure 3 presents the observed event model that could be used in monitoring medication adherence.

Time represents the timestamp that can be associated with an observed event. An observed event can have a *start date* and an *end date* (same timestamp for instantaneous events) and sometimes a reported time (event reported after event-end date). Usually an event refers to a specific *entity* (patient, object, location); *data* represents the *value* associated to the event and its *confidence level*. The confidence level can be specialized into specific uncertainty representation (probability, possibility, qualitative). *Source* represents the event's source (observation, report) and the used method (from sensor, from processing component, from manual input). This event model is used to define the event taxonomy, which is used to listen to specific event types. Events can be viewed on three aspects (Fig. 4), which provide three event classification schemes (source-based, contextual information based and related entity based).

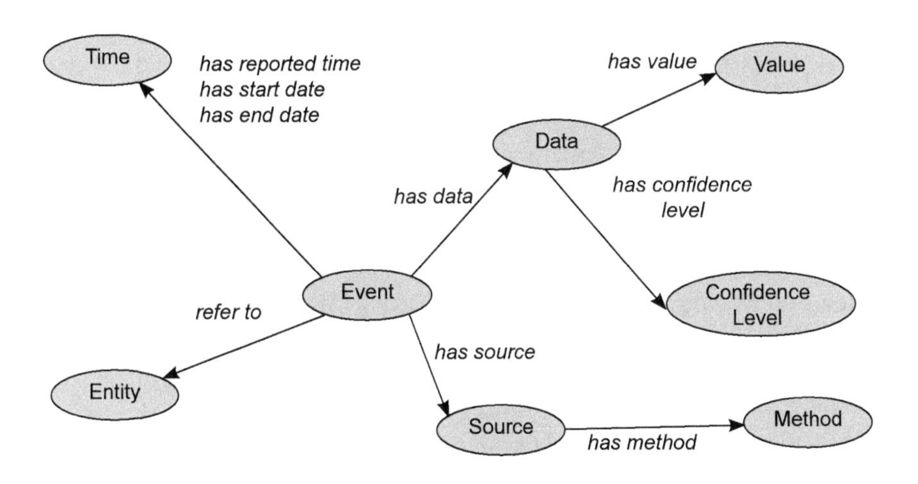

Fig. 3. Modelling events for monitoring medication adherence.

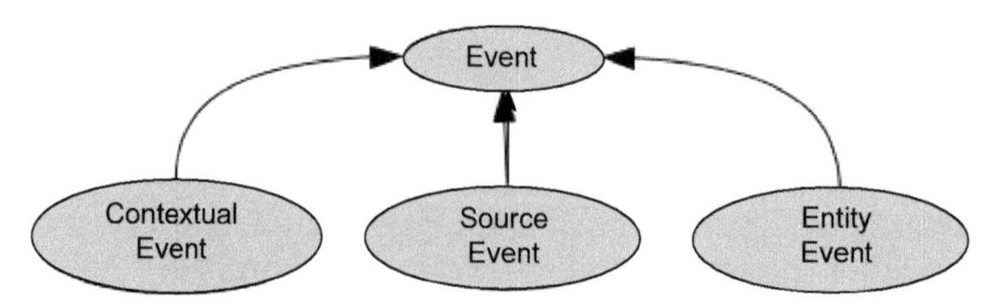

Fig. 4. High-level event taxonomy.

The *source* aspect (Fig. 5) classifies events according to the *source* type and the used *method. Source type* can be *observation* (raw, inferred and manual) and *report* (questionnaire, assessment summary); *method type* can be *sensor* (aggregated sensor, specific sensor type), *processing component* (fall detection, health issue detection, human activity recognition), and *manual input* (user input, aggregator input, medical staff input). For instance, an accelerometer event can be classified as a *raw observation* from an *accelerometer sensor* (sensor-specific type).

The *contextual* aspect (Fig. 6) classifies events according to the provided contextual information (motion, location, measured property, situation). *Motion* event can be, for instance, an *acceleration*, a *relative change* (floor, altitude and step changes) and a *motion state* (stationary, moving, cycling, walking, running, automotive). *Location* event can be *topological* (inside, outside, enter, leave a region), *proximity* (near, immediate, at, far) and *relative position* (left side, behind). *Measured property* can be *ambient* (temperature, pressure, luminosity, humidity, magnetic field) and *physiological* (heart rate, blood pressure, oxygen saturation). *Situation* event can be *smart environment state* event (automation device state, service state), *health* event (mental health, social health, physiological health), *health issue* event and *human activity* event (activity, task, action, gesture, activity of daily living).

The *entity* aspect (Fig. 7) classifies events according to the entity related to the event. *Entity* event can be a *person* event (user, patient, relative, friend, caregiver, aggregator, medical staff), *location entity* event (point, place, architectural, domicile, room, geographic region, geofence region, beacon region) and *abstract entity* event (smartphone system, web service, mobile patient diary, agent).

Thus, the observed '*taking medication activity*' event can be classified as an inferred observation from activity recognition event (source aspect), as an activity event (contextual aspect) and as a patient event (entity aspect).

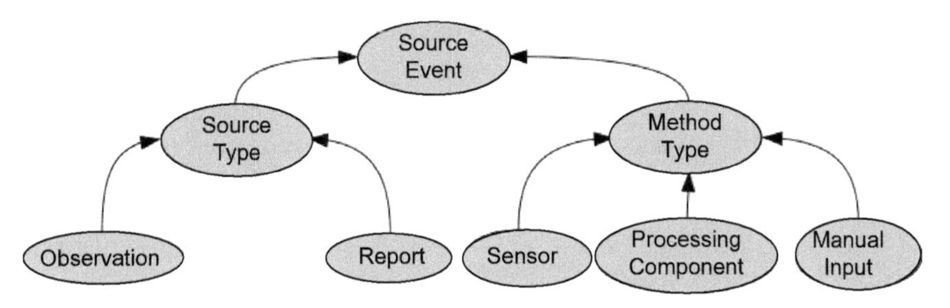

Fig. 5. High-level taxonomy for source aspect.

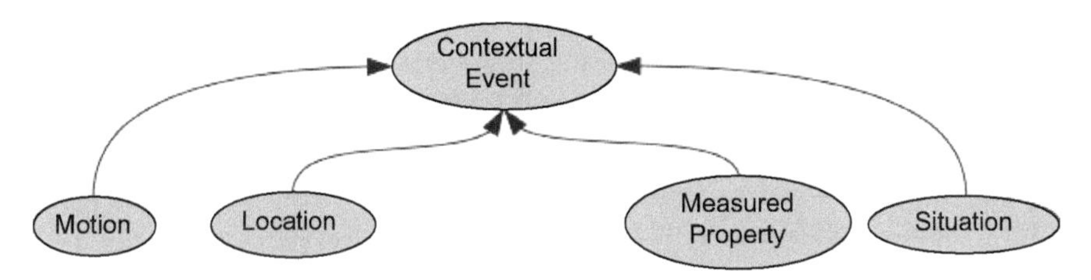

Fig. 6. High-level taxonomy for contextual aspect.

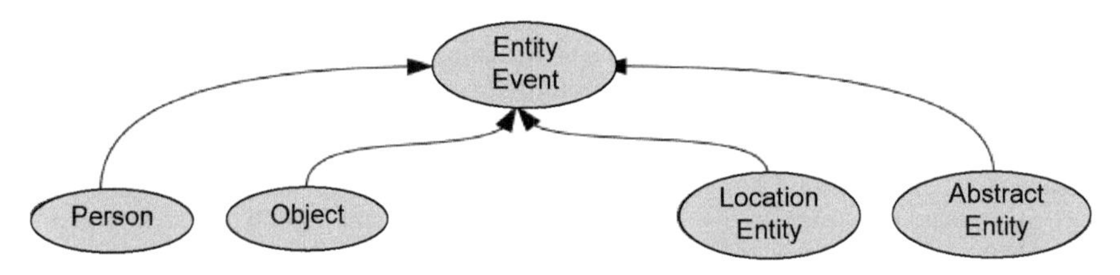

Fig. 7. High-level taxonomy for entity aspect.

Activity Recognition

Activity recognition can be restricted to activities related to the on-going personalized self-management plan. Using an ontological-driven activity-recognition approach allows one to instantiate relevant activity models. Depending on the available event sources and activities' properties, events are selected and fused to infer high-level contexts (activity, task, action, etc.). It should be noted that the activity modelling and the process of segmentation, selection and fusion of events are still under investigation. Activity modelling and the reasoning process must take into account uncertainty and complex temporal relations.

5. Conclusion

Patients following self-management programs can benefit from using smart environments' services, such as activity recognition and assistance services. Interactions between self-management program and smart environment can improve self-management strategies to overcome barriers related to a patient's goals (medication adherence) in self-managing his disease. Ontological-driven activity recognition and self-management can interact by using low (event) to high (activity) context information. An event model and taxonomy, which take into account events that could occur in the context of monitoring a self-efficacy strategy, was presented. Our ongoing work focuses on activity modelling and reasoning, which must handle uncertainty, imperfect information and complex temporal relations.

Keywords: Self-management programs, activity recognition, self-care, self-efficacy, medication adherence, smart environment, ambient assisted living, health behavior, social cognitive theory, ontology

References

Abidi, Samina Raza, Syed Sibte Raza Abidi and Ashraf Abusharek. 2013. A semantic web-based mobile framework for designing personalized patient self-management interventions. pp. 1–4. *In*: Proceedings of the 1st Conference on Mobile and Information Technologies in Medicine. Prague, Czech Republic: Czech Technical University, Prague.

Abidi, Syed Sibte Raza and Samina Abidi. 2013. An ontology-driven personalization framework for designing theory-driven self-management interventions. *In*: David Riaño, Richard Lenz, Silvia Miksch, Mor Peleg, Manfred Reichert and Annette ten Teije (Eds.). Process Support and Knowledge Representation in Health Care, 8268: 97–112. Lecture Notes in Computer Science. Switzerland: Springer International Publishing. doi:10.1007/978-3-319-03916-9_8.

Aitken, Murray and Silvia Valkova. 2013. Avoidable costs in US healthcare: The $200 Billion Opportunity for Using Medicines More Responsibly. IMS Institute for Healthcare Informatics, Parsippany, NJ.

Aljasem, L.I., M. Peyrot, L. Wissow and R.R. Rubin. 2001. The Impact of Barriers and Self-Efficacy on Self-Care Behaviors in Type 2 Diabetes. The Diabetes Educator. 27(3): 393–404. doi:10.1177/014572170102700309.

Allen, James F. and George Ferguson. 1994. Actions and events in interval temporal logic. Journal of Logic and Computation. 4(5): 531–79. doi:10.1093/logcom/4.5.531.

Aloulou, Hamdi, Mounir Mokhtari, Thibaut Tiberghien, Romain Endelin and Jit Biswas. 2015. Uncertainty Handling in Semantic Reasoning for Accurate Context Understanding. Knowledge-Based Systems. 77 (March): 16–28. doi:10.1016/j.knosys.2014.12.025.

Atreja, Ashish, Naresh Bellam and Susan R. Levy. 2005. Strategies to enhance patient adherence: making it simple. MedGenMed: Medscape General Medicine. 7(1): 4.

Baader, Franz, Ian Horrocks and Ulrike Sattler. 2004. Description logics. pp. 3–28. *In*: P.D. Dr. Steffen Staab and Professor Dr. Rudi Studer (Eds.). Handbook on Ontologies. International Handbooks on Information Systems. Springer Berlin Heidelberg.

Bandura, Albert. 2001. Social Cognitive Theory of Mass Communication. Media Psychology 3(3). Lawrence Erlbaum Associates, Inc.: 265–99. doi:10.1207/S1532785XMEP0303_03.
Bandura, Albert. 2004. Health Promotion by Social Cognitive Means. Health Education & Behavior : The Official Publication of the Society for Public Health Education. 31(2): 143–64. doi:10.1177/1090198104263660.
Bardram, Jakob E., Mads Frost, Károly Szántó and Gabriela Marcu. 2012. The MONARCA self-assessment system: a persuasive personal monitoring system for bipolar patients. pp. 21–30. *In*: Proceedings of the 2Nd ACM SIGHIT International Health Informatics Symposium. IHI '12. New York, NY, USA: ACM. doi:10.1145/2110363.2110370.
Bender, B., H. Milgrom and C. Rand. 1997. Nonadherence in asthmatic patients: is there a solution to the problem? Annals of Allergy, Asthma & Immunology: Official Publication of the American College of Allergy, Asthma, & Immunology. 79(3): 177–85; quiz 185–86. doi:10.1016/S1081-1206(10)63001-3.
Bilinski, Piotr and Francois Bremond. 2011. Evaluation of local descriptors for action recognition in videos. pp. 61–70. *In*: James L. Crowley, Bruce A. Draper and Monique Thonnat (Eds.). Computer Vision Systems. Lecture Notes in Computer Science. Springer Berlin Heidelberg.
Bodenheimer, Thomas, Kate Lorig, Halsted Holman and Kevin Grumbach. 2002. Patient self-management of chronic disease in primary care. JAMA : The Journal of the American Medical Association. 288(19). American Medical Association: 2469–75. doi:10.1001/jama.288.19.2469.
Boudreaux, Edwin D., Jennifer L. Francis, Cindy L. Carmack Taylor, Isabel C. Scarinci and Phillip J. Brantley. 2003. Changing multiple health behaviors: Smoking and exercise. Preventive Medicine. 36(4): 471–78. doi:10.1016/S0091-7435(02)00048-8.
Burgoon, Judee K., Michael Pfau, Roxanne Parrott, Thomas Birk, Ray Coker and Michael Burgoon. 1987. Relational Communication, Satisfaction, Compliance-gaining Strategies, and Compliance in Communication between Physicians and Patients. Communication Monographs. 54(3). Taylor & Francis Group: 307–24. doi:10.1080/03637758709390235.
Chan, Marie, Daniel Estève, Jean-Yves Fourniols, Christophe Escriba and Eric Campo. 2012. Smart wearable systems: Current status and future challenges. Artificial Intelligence in Medicine. 56(3): 137–56. doi:10.1016/j.artmed.2012.09.003.
Chen, Liming, C.D. Nugent and Hui Wang. 2012. A knowledge-driven approach to activity recognition in smart homes. IEEE Transactions on Knowledge and Data Engineering. 24(6): 961–74. doi:10.1109/TKDE.2011.51.
Chen, Liming, Chris Nugent, Maurice Mulvenna, Dewar Finlay, Xin Hong and Michael Poland. 2008. Using event calculus for behaviour reasoning and assistance in a smart home. pp. 81–89. *In*: Sumi Helal, Simanta Mitra, Johnny Wong, Carl K. Chang and Mounir Mokhtari (Eds.). Smart Homes and Health Telematics. Lecture Notes in Computer Science. Springer Berlin Heidelberg.
Christensen, Henrik Bærbak. 2002. Using logic programming to detect activities in pervasive healthcare. pp. 421–36. *In*: Peter J. Stuckey (Ed.). Logic Programming. Lecture Notes in Computer Science. Springer Berlin Heidelberg.
Clark, N.M. and J.A. Dodge. 1999. Exploring self-efficacy as a predictor of disease management. Health Education & Behavior: The Official Publication of the Society for Public Health Education. 26(1): 72–89.
Cohen, Philip R., C. Raymond Perrault and James F. Allen. 1982. Beyond question answering. pp. 245–74. *In*: Wendy G. Lehnert and Martin H. Ringle (Eds.). Strategies for Natural Language Processing. Lawrence Erlbaum Associates.
Compton, Michael, Payam Barnaghi, Luis Bermudez, Raúl García-Castro, Oscar Corcho, Simon Cox, John Graybeal et al. 2012. The SSN Ontology of the W3C Semantic Sensor Network Incubator Group. Web Semantics: Science, Services and Agents on the World Wide Web 17 (December): 25–32. doi:10.1016/j.websem.2012.05.003.

Cook, Diane J. and Narayanan Krishnan. 2014. Mining the home environment. Journal of Intelligent Information Systems. 43(3): 503–19. doi:10.1007/s10844-014-0341-4.
Cramer, Joyce A. 1998. Enhancing patient compliance in the elderly. Drugs & Aging. 12(1): 7–15. doi:10.2165/00002512-199812010-00002.
Daltroy, L.H., J.N. Katz, C.I. Morlino and M.H. Liang. 1991. Improving doctor patient communication. Psychiatr Med. 2: 31–35.
Delachaux, Benoît, Julien Rebetez, Andres Perez-Uribe and Héctor Fabio Satizábal Mejia. 2013. Indoor activity recognition by combining one-vs.-all neural network classifiers exploiting wearable and depth sensors. *In*: Ignacio Rojas, Gonzalo Joya and Joan Cabestany (Eds.). Advances in Computational Intelligence 12th International Work Conference on Artificial Neural Networks, IWANN 2013, Puerto de La Cruz, Tenerife, Spain, June 12–14, 2013, Proceedings, Part II. 7903: 216–23. Lecture Notes in Computer Science. Berlin, Heidelberg: Springer Berlin Heidelberg. doi: 10.1007/978-3-642-38682-4.
DiMatteo, M. Robin. 2004. Variations in Patients' Adherence to medical recommendations: a quantitative review of 50 years of research. Medical Care. 42(3): 200–209.
Dubois, Didier and Henri Prade. 2006. Possibility theory and its applications: a retrospective and prospective view. pp. 89–109. *In*: Giacomo Della Riccia, Didier Dubois, Rudolf Kruse, and Hanz-Joachim Lenz (Eds.). Decision Theory and Multi-Agent Planning. CISM International Centre for Mechanical Sciences. Springer Vienna.
Eraker, S.A., J.P. Kirscht and M.H. Becker. 1984. Understanding and improving patient compliance. Annals of Internal Medicine. 100(2): 258–68.
Friedman, Nir, Kevin Murphy and Stuart Russell. 1998. Learning the structure of dynamic probabilistic networks. pp. 139–47. *In*: Proceedings of the Fourteenth Conference on Uncertainty in Artificial Intelligence. UAI'98. San Francisco, CA, USA: Morgan Kaufmann Publishers Inc.
Gangemi, Aldo and Peter Mika. 2003. Understanding the semantic web through descriptions and situations. pp. 689–706. *In*: Robert Meersman, Zahir Tari and Douglas C. Schmidt (Eds.). On The Move to Meaningful Internet Systems 2003: CoopIS, DOA, and ODBASE. Lecture Notes in Computer Science. Springer Berlin Heidelberg.
Geib, C. 2006. Plan recognition. pp. 77–100. *In*: A. Kott and W.M. McEneaney (Eds.). Adversarial Reasoning: Computational Approaches to Reading the Opponent's Mind. Chapman & Hall/CRC.
Gillani Fahad, Labiba, Arshad Ali and Muttukrishnan Rajarajan. 2015. Learning models for activity recognition in smart homes. *In*: Kuinam J. Kim (Ed.). Information Science and Applications. 339: 819–26. Lecture Notes in Electrical Engineering. Springer Berlin Heidelberg. doi:10.1007/978-3-662-46578-3_97.
Guidoux, Romain, Martine Duclos, Gérard Fleury, Philippe Lacomme, Nicolas Lamaudière, Pierre-Henri Manenq, Ludivine Paris, Libo Ren and Sylvie Rousset. 2014. A smartphone-driven methodology for estimating physical activities and energy expenditure in free living conditions. Journal of Biomedical Informatics, Special Section: Methods in Clinical Research Informatics. 52 (December): 271–78. doi:10.1016/j.jbi.2014.07.009.
Haakstad, Lene A.H., Ingvild Gundersen and Kari Bø. 2010. Self-reporting compared to motion monitor in the measurement of physical activity during pregnancy. Acta Obstetricia et Gynecologica Scandinavica. 89(6): 749–56. doi:10.3109/00016349.2010.484482.
Hall, J.A., D.L. Roter and N.R. Katz. 1988. Meta-analysis of correlates of provider behavior in medical encounters. Medical Care. 26(7): 657–75.
Hayes, Pat, Jeremy Carroll, Chris Welty, Michael Uschold, Bernard Vatant, Frank Manola, Ivan Herman and Jamie Lawrence. 2006. Defining N-Ary Relations on the Semantic Web. W3C Working Group Note. http://www.w3.org/TR/swbp-n-aryRelations/.
Haynes, R.B., E. Ackloo, N. Sahota, H.P. McDonald and X. Yao. 2008. Interventions for Enhancing Medication Adherence. The Cochrane Database of Systematic Reviews, No. 2 (January): CD000011. doi:10.1002/14651858.CD000011.pub3.

Haynes, R.B., H. McDonald, A.X. Garg and P. Montague. 2002. Interventions for Helping Patients to Follow Prescriptions for Medications. The Cochrane Database of Systematic Reviews, No. 2 (January): CD000011. doi:10.1002/14651858.CD000011.
Helaoui, Rim, Daniele Riboni and Heiner Stuckenschmidt. 2013. A probabilistic ontological framework for the recognition of multilevel human activities. pp. 345–54. *In*: Proceedings of the 2013 ACM International Joint Conference on Pervasive and Ubiquitous Computing. UbiComp '13. New York, NY, USA: ACM. doi:10.1145/2493432.2493501.
Horrocks, Ian, Peter F. Patel-Schneider, Deborah L. McGuinness and Christopher A. Welty. 2007. OWL: A description-logic-based ontology language for the semantic web. pp. 458–86. *In*: Franz Baader, Diego Calvanese, Deborah L. McGuinness, Daniele Nardi and Peter F. Patel-Schneider (Eds.). The Description Logic Handbook: Theory, Implementation and Applications. Cambridge University Press.
Horvitz, Eric, Jack Breese, David Heckerman, David Hovel and Koos Rommelse. 1998. The Lumière Project: Bayesian user modeling for inferring the goals and needs of software users. pp. 256–65. *In*: UAI'98 Proceedings of the Fourteenth Conference on Uncertainty in Artificial Intelligence, Morgan Kaufmann Publishers Inc.
Hu, Ninghang, G. Englebienne, Zhongyu Lou and B. Krose. 2014. Learning latent structure for activity recognition. pp. 1048–53. *In*: 2014 IEEE International Conference on Robotics and Automation (ICRA), doi:10.1109/ICRA.2014.6906983.
Kadouche, Rachid, Belkacem Chikhaoui and Bessam Abdulrazak. 2010. User's behavior study for smart houses occupant prediction. Annales Des Telecommunications/Annals of Telecommunications. 65(9-10): 539–43. doi:10.1007/s12243-010-0166-2.
Katz, J.R. 1997. Back to basics: providing effective patient teaching. The American Journal of Nursing. 97(5): 33–36.
Katz, S., A.B. Ford, R.W. Moskowitz, B.A. Jackson and M.W. Jaffe. 1963. Studies of illness in the aged: the index of ADL: A Standardized Measure of Biological and Psychosocial Function. JAMA. 185(12): 914–19.
Kautz, Henry A. 1991. A formal theory of plan recognition and its implementation. pp. 69–124. *In*: Ronald J. Brachman, James F. Allen, Henry A. Kautz, Richard N. Pelavin and Josh D. Tenenberg (Eds.). Reasoning about Plans. San Francisco, CA, USA: Morgan Kaufmann Publishers Inc.
Knublauch, Holger, James A. Hendler and Kingsley Idehen. 2011. SPIN—Overview and Motivation. http://www.w3.org/Submission/spin-Overview/.
Kronish, Ian M. and Siqin Ye. 2013. Adherence to cardiovascular medications: lessons learned and future directions. Progress in Cardiovascular Diseases. 55(6): 590–600. doi:10.1016/j.pcad.2013.02.001.
Meditskos, Georgios, Stamatia Dasiopoulou, Vasiliki Efstathiou and Ioannis Kompatsiaris. 2013a. Ontology patterns for complex activity modelling. pp. 144–57. *In*: Leora Morgenstern, Petros Stefaneas, François Lévy, Adam Wyner and Adrian Paschke (Eds.). Theory, Practice, and Applications of Rules on the Web. Lecture Notes in Computer Science. Springer Berlin Heidelberg.
Meditskos, Georgios, Stamatia Dasiopoulou, Vasiliki Efstathiou and Ioannis Kompatsiaris. 2013b. SP-ACT: A Hybrid Framework for Complex Activity Recognition Combining OWL and SPARQL Rules. pp. 25–30. *In*: 2013 IEEE International Conference on Pervasive Computing and Communications Workshops (PERCOM Workshops). doi:10.1109/PerComW.2013.6529451.
Meditskos, Georgios, Stamatia Dasiopoulou and Ioannis Kompatsiaris. 2015. MetaQ: A Knowledge-Driven Framework for Context-Aware Activity Recognition Combining SPARQL and OWL 2 Activity Patterns. Pervasive and Mobile Computing (in press). doi:10.1016/j.pmcj.2015.01.007.

Milgrom, Henry, Bruce Bender, Lynn Ackerson, Pamela Bowry, Bernita Smith and Cynthia Rand. 1996. Noncompliance and treatment failure in children with asthma. Journal of Allergy and Clinical Immunology. 98(6): 1051–57. doi:10.1016/S0091-6749(96)80190-4.

Morningstar, B.A., I.S. Sketris, G.C. Kephart and D.A. Sclar. 2002. Variation in pharmacy prescription refill adherence measures by type of oral antihyperglycaemic drug therapy in seniors in Nova Scotia, Canada. Journal of Clinical Pharmacy and Therapeutics. 27(3): 213–20. doi:10.1046/j.1365-2710.2002.00411.x.

Morrow, D. 1997. Improving Consultations between Health-Care Professionals and Older Clients: Implications for Pharmacists. International Journal of Aging & Human Development. 44(1): 47–72.

Nazerfard, Ehsan, Barnan Das, Lawrence B. Holder and Diane J. Cook. 2010. Conditional random fields for activity recognition in smart environments. pp. 282–86. *In*: Proceedings of the 1st ACM International Health Informatics Symposium, IHI '10. New York, NY, USA: ACM. doi:10.1145/1882992.1883032.

Niepert, Mathias, Jan Noessner and Heiner Stuckenschmidt. 2011. Log-linear description logics. pp. 2153–58. *In*: IJCAI International Joint Conference on Artificial Intelligence. AAAI Press. doi:10.5591/978-1-57735-516-8/IJCAI11-359.

Oka, Roberta K., Teresa DeMarco and William L. Haskell. 2005. Effect of treadmill testing and exercise training on self-efficacy in patients with heart failure. European Journal of Cardiovascular Nursing: Journal of the Working Group on Cardiovascular Nursing of the European Society of Cardiology. 4(3): 215–19. doi:10.1016/j.ejcnurse.2005.04.004.

Okeyo, George, Liming Chen and Hui Wang. 2014. Combining ontological and temporal formalisms for composite activity modelling and recognition in smart homes. Future Generation Computer Systems, Special Issue on Ubiquitous Computing and Future Communication Systems. 39 (October): 29–43. doi:10.1016/j.future.2014.02.014.

Okeyo, George, Liming Chen, Hui Wang and Roy Sterritt. 2014. Dynamic sensor data segmentation for real-time knowledge-driven activity recognition. Pervasive and Mobile Computing 10, Part B (February): 155–72. doi:10.1016/j.pmcj.2012.11.004.

Osterberg, Lars and Terrence Blaschke. 2005. Adherence to medication. New England Journal of Medicine. 353(5). Massachusetts Medical Society: 487–97. doi:10.1056/NEJMra050100.

Patterson, D.J., D. Fox, H. Kautz and M. Philipose. 2005. Fine-grained activity recognition by aggregating abstract object usage. pp. 44–51. *In*: Ninth IEEE International Symposium on Wearable Computers, 2005. Proceedings. doi:10.1109/ISWC.2005.22.

Patterson, Timothy, Ian Cleland, Phillip J. Hartin, Chris D. Nugent, Norman D. Black, Mark P. Donnelly, Paul J. McCullagh, Huiru Zheng and Suzanne McDonough. 2015. Home-based self-management of dementia: closing the loop. *In*: Antoine Geissbühler, Jacques Demongeot, Mounir Mokhtari, Bessam Abdulrazak and Hamdi Aloulou (Eds.). Inclusive Smart Cities and E-Health. 9102: 232–43. Lecture Notes in Computer Science. Cham: Springer International Publishing. doi: 10.1007/978-3-319-19312-0.

Peterson, Andrew M., Liza Takiya and Rebecca Finley. 2003. Meta-analysis of trials of interventions to improve medication adherence. American Journal of Health-System Pharmacy : AJHP, Official Journal of the American Society of Health-System Pharmacists. 60(7): 657–65.

Phua, C., V.S.F. Foo, J. Biswas, A. Tolstikov, Aung-Phyo-Wai Aung, J. Maniyeri, Weimin Huang, Mon-Htwe That, Duangui Xu and A.K.W. Chu. 2009. 2-Layer erroneous-plan recognition for dementia patients in smart homes. pp. 21–28. *In*: 11th International Conference on E-Health Networking, Applications and Services, 2009. Healthcom 2009. doi:10.1109/HEALTH.2009.5406183.

Quinn, Catherine, Gill Toms, Daniel Anderson and Linda Clare. 2015. A review of self-management interventions for people with dementia and mild cognitive impairment. Journal of Applied Gerontology, January. doi:10.1177/0733464814566852.

Rabiner, L. 1989. A Tutorial on Hidden Markov Models and Selected Applications in Speech Recognition. Proceedings of the IEEE. 77(2): 257–86. doi:10.1109/5.18626.
Rosenberg, E.E., M.T. Lussier and C. Beaudoin. 1997. Lessons for clinicians from physician-patient communication literature. Archives of Family Medicine. 6(3): 279–83.
Roter, D.L., J.A. Hall, R. Merisca, B. Nordstrom, D. Cretin and B. Svarstad. 1998. Effectiveness of interventions to improve patient compliance: a meta-analysis. Medical Care. 36(8): 1138–61.
Roy, Patrice C., Newres Al Haider, William Van Woensel, Ahmad Marwan Ahmad and Syed S.R. Abidi. 2014. Towards guideline compliant clinical decision support system integration in smart and mobile environments: formalizing and using clinical guidelines for diagnosing sleep apnea. pp. 38–43. *In*: Bruno Bouchard, Abdenour Bouzouane, Sylvain Giroux, Alex Mihailidis and Sébastien Guillet (Eds.). AAAI-14 Workshop on Artificial Intelligence Applied to Assistive Technologies and Smart Environments.
Roy, Patrice C., Bruno Bouchard, Abdenour Bouzouane and Sylvain Giroux. 2009. A hybrid plan recognition model for Alzheimer's patients: interleaved-erroneous dilemma. Web Intelligence and Agent Systems. 7(4): 375–97. doi:10.3233/WIA-2009-0175.
Roy, Patrice C., Abdenour Bouzouane, Sylvain Giroux and Bruno Bouchard. 2011. Possibilistic activity recognition in smart homes for cognitively impaired people. Applied Artificial Intelligence. 25(10): 883–926. doi:10.1080/08839514.2011.617248.
Shafer, G. 1976. A Mathematical Theory of Evidence. Princeton University Press Princeton, NJ.
Stankovski, Vlado and Jernej Trnkoczy. 2006. Application of decision trees to smart homes. pp. 132–45. *In*: Juan Carlos Augusto and Chris D. Nugent (Eds.). Designing Smart Homes. Lecture Notes in Computer Science. Springer Berlin Heidelberg.
Tapia, Emmanuel Munguia, Stephen S. Intille and Kent Larson. 2004. Activity recognition in the home using simple and ubiquitous sensors. pp. 158–75. *In*: Alois Ferscha and Friedemann Mattern (Eds.). Pervasive Computing. Lecture Notes in Computer Science. Springer Berlin Heidelberg.
Weinland, Daniel, Remi Ronfard and Edmond Boyer. 2011. A survey of vision-based methods for action representation, segmentation and recognition. Computer Vision and Image Understanding. 115(2): 224–41. doi:10.1016/j.cviu.2010.10.002.
WHO. 2003. Adherence to long-term therapies: evidence for action. World Health Organization, Geneva, Switzerland.
Ye, Juan, Graeme Stevenson, Simon Dobson, Michael O'Grady and Gregory O'Hare. 2013. Perceiving and interpreting smart home datasets with PI. Journal of Ambient Intelligence and Humanized Computing. 4(6): 717–29. doi:10.1007/s12652-012-0148-5.

Index

An environmentally friendly book printed and bound in England by www.printondemand-worldwide.com

PEFC Certified

This product is from sustainably managed forests and controlled sources

www.pefc.org

PEFC

PEFC/16-33-415

This book is made of chain-of-custody materials; FSC materials for the cover and PEFC materials for the text pages.